THE INTEGRATION OF THE Humanities and Arts WITH Sciences, Engineering, and Medicine IN HIGHER EDUCATION

Branches FROM THE Same Tree

David Skorton and Ashley Bear, *Editors*

Committee on Integrating Higher Education in the Arts, Humanities, Sciences, Engineering, and Medicine

Board on Higher Education and Workforce

Policy and Global Affairs

A Consensus Study Report of
The National Academies of
SCIENCES • ENGINEERING • MEDICINE

THE NATIONAL ACADEMIES PRESS
Washington, DC
www.nap.edu

THE NATIONAL ACADEMIES PRESS   500 Fifth Street, NW   Washington, DC 20001

This activity was supported by contracts between the National Academy of Sciences and the Andrew W. Mellon Foundation (#11600619), the National Endowment for the Humanities and the National Endowment for the Arts (#AH253080-16), the National Academy of Sciences Scientists and Engineers for the Future Fund, and the Teagle Foundation. Any opinions, findings, conclusions, or recommendations expressed in this publication do not necessarily reflect the views of any organization or agency that provided support for the project.

International Standard Book Number-13: 978-0-309-47061-2
International Standard Book Number-10: 0-309-47061-7
Digital Object Identifier: https://doi.org/10.17226/24988
Library of Congress Control Number: 2018941713

Additional copies of this publication are available for sale from the National Academies Press, 500 Fifth Street, NW, Keck 360, Washington, DC 20001; (800) 624-6242 or (202) 334-3313; http://www.nap.edu.

Printed in the United States of America

Suggested citation: National Academies of Sciences, Engineering, and Medicine. 2018. *The Integration of the Humanities and Arts with Sciences, Engineering, and Medicine in Higher Education: Branches from the Same Tree*. Washington, DC: The National Academies Press. doi: https://doi.org/10.17226/24988.

The National Academies of

SCIENCES • ENGINEERING • MEDICINE

The **National Academy of Sciences** was established in 1863 by an Act of Congress, signed by President Lincoln, as a private, nongovernmental institution to advise the nation on issues related to science and technology. Members are elected by their peers for outstanding contributions to research. Dr. Marcia McNutt is president.

The **National Academy of Engineering** was established in 1964 under the charter of the National Academy of Sciences to bring the practices of engineering to advising the nation. Members are elected by their peers for extraordinary contributions to engineering. Dr. C. D. Mote, Jr., is president.

The **National Academy of Medicine** (formerly the Institute of Medicine) was established in 1970 under the charter of the National Academy of Sciences to advise the nation on medical and health issues. Members are elected by their peers for distinguished contributions to medicine and health. Dr. Victor J. Dzau is president.

The three Academies work together as the **National Academies of Sciences, Engineering, and Medicine** to provide independent, objective analysis and advice to the nation and conduct other activities to solve complex problems and inform public policy decisions. The National Academies also encourage education and research, recognize outstanding contributions to knowledge, and increase public understanding in matters of science, engineering, and medicine.

Learn more about the National Academies of Sciences, Engineering, and Medicine at **www.nationalacademies.org**.

COMMITTEE ON INTEGRATING HIGHER EDUCATION IN THE ARTS, HUMANITIES, SCIENCES, ENGINEERING, AND MEDICINE[1]

DAVID SKORTON [NAM] (*Chair*), Secretary, Smithsonian Institution
SUSAN ALBERTINE, Senior Scholar, Association of American Colleges & Universities
NORMAN AUGUSTINE (NAS/NAE), Retired Chairman and CEO, Lockheed Martin Corporation
LAURIE BAEFSKY, Executive Director, Arts Engine and the Alliance for the Arts in Research Universities (a2ru), University of Michigan
KRISTIN BOUDREAU, The Paris and Fletcher Distinguished Professor of Humanities, Department Head, Humanities and Arts, Worcester Polytechnic Institute
NORMAN BRADBURN, Senior Fellow, NORC, The Tiffany and Margaret Blake Distinguished Service Professor Emeritus, The University of Chicago
AL BUNSHAFT, Senior Vice President, Global Affairs and Workforce of the Future, Dassault Systèmes' Americas
GAIL BURD, Senior Vice Provost for Academic Affairs, Distinguished Professor, Molecular and Cellular Biology, Cellular and Molecular Medicine, University of Arizona
EDWARD DERRICK, Independent Consultant
E. THOMAS EWING, Professor of History, Associate Dean, Graduate Studies, Research, and Diversity, The College of Liberal Arts and Human Sciences, Virginia Tech
J. BENJAMIN HURLBUT, Associate Professor of Biology and Society, The School of Life Sciences, Arizona State University
PAMELA L. JENNINGS, Professor and Head, Department of Art + Design, College of Design, North Carolina State University
YOUNGMOO KIM, Director, The Expressive and Creative Interaction Technologies (ExCITe) Center, Professor of Electrical and Computer Engineering, Drexel University
ROBERT MARTELLO, Associate Dean for Curriculum and Academic Programs, Professor of the History of Science and Technology, Olin College
GUNALAN NADARAJAN, Dean and Professor, The Penny W. Stamps School of Art and Design, The University of Michigan

[1] Paul Bevilaqua (NAE), Retired Manager, Advanced Development Programs, Lockheed Martin Aeronautics Company, resigned from the committee in November 2017.

THOMAS F. NELSON LAIRD, Associate Professor, Higher Education and Student Affairs Program, and Director, Center for Postsecondary Research, Indiana University Bloomington
LYNN PASQUERELLA, President, The Association of American Colleges & Universities
SUZANNA ROSE, Founding Associate Provost, Office to Advance Women, Equity, and Diversity, Professor of Psychology and Women's and Gender Studies, Florida International University
BONNIE THORNTON DILL, Dean, College of Arts and Humanities and Professor of Women's Studies, The University of Maryland
LAURA VOSEJPKA, Founding Dean, College of Sciences and Liberal Arts, Kettering University
LISA M. WONG, Co-Director, The Arts and Humanities Initiative, Assistant Clinical Professor of Pediatrics, Harvard Medical School

Study Staff

ASHLEY BEAR, Study Director
AUSTEN APPLEGATE, Senior Program Assistant
ADRIANA COUREMBIS, Financial Officer
ELIZABETH GARBEE, Christine Mirzayan Science and Technology Policy Graduate Fellow
KELLYANN JONES-JAMTGAARD, Christine Mirzayan Science and Technology Policy Graduate Fellow
JAY LABOV, Senior Advisor for Education and Communication
IRENE NGUN, Research Associate
THOMAS RUDIN, Director, Board on Higher Education and Workforce
J. D. TALASEK, Director of Cultural Programs

Consultants

STEVE OLSON, Writer
MATTHEW MAYHEW, William Ray and Marie Adamson Flesher Professor of Educational Administration, The Ohio State University
HANNAH STEWART-GAMBINO, Professor of Government & Law and International Affairs, Lafayette College
JENNIFER STROUD ROSSMANN, Professor of Mechanical Engineering, Lafayette College

BOARD ON HIGHER EDUCATION AND WORKFORCE

Preface

American higher education has for generations been the envy of the world. Whether because of the enormous output of research, scholarship, and creative activity or the great diversity of offerings—running the gamut from community colleges to liberal arts colleges, research universities, conservatories, technical schools, and many other categories—American colleges and universities are widely admired and emulated across the globe.

In tracing the history of American higher education, we find much to be proud of, but we also see over the past few decades a growing tension between the broad and integrated education commonly referred to as liberal education and the increasing specialization in higher education as individual disciplines and administrative structures drive a fragmentation of curricula. This tension between broad, integrated education and specialized, disciplinary studies has heightened during periods of economic challenge, particularly since the Great Recession that began in 2008. Students and parents increasingly have focused their aspirations and plans on a vocationally driven approach, emphasizing fields where immediate post-graduation employment seems more certain and more remunerative.

Ironically, as this movement toward narrower, disciplinary education has progressed inexorably, many employers—even, and, in fact, especially in "high tech" areas—have emphasized that learning outcomes associated with integrated education, such as critical thinking, communication, teamwork, and abilities for lifelong learning, are more, not less, desirable. With the enormous strides in technology, including artificial intelligence, machine learning, robotics, and communications, graduates will need such transferable and uniquely human skills to be able to adaptively and continuously learn to work with, and alongside, new technologies. Further, each person entering the job market today will look forward not only to several jobs, but also several careers, during her working life. All of these factors have led to the expectation that current generations entering the workforce may, for the first time in recent American history, face a more uncertain future than their parents' generation.

Faculty and administrators, who are concerned that an education focused on a single discipline will not best prepare graduates for the challenges and opportunities presented by work, life, and citizenship in the 21st century, are advocating for an approach to education that moves beyond the general education requirements found at almost all institutions, to an approach to higher education that intentionally integrates knowledge in the arts, humanities, physical and life sciences, social sciences, engineering, technology, mathematics, and the biomedical disciplines. In this approach, which we refer to in this report simply as "integration," professors help students understand the connections among the disciplines and emphasize the point made by Einstein that all disciplines and forms of inquiry are "branches from the same tree." Extending this metaphor, advocates of integration see all human knowledge as both fundamentally connected, a network of branches arising from a trunk made up of human curiosity, passion, and drive, but also generative, as new branches split off and grow from old branches, extending into new spaces or coming in contact with other branches in new ways.

Against this backdrop, the Board on Higher Education and Workforce (BHEW) of the National Academies of Sciences, Engineering, and Medicine conducted a study focused on better understanding the impact of an integrated educational approach on students. Specifically, the committee was charged with "*examining the evidence behind the assertion that educational programs that mutually integrate learning experiences in the humanities and arts with science, technology, engineering, math, and medicine (STEMM) lead to improved educational and career outcomes for undergraduate and graduate students.*" To be clear, our task was neither to reject the disciplines, which this committee sees as vital sources of expertise, creativity, and innovation, nor to argue that an integrative approach is superior to more established models of general education. Rather, our task was to examine what the existing evidence can tell us about the impact on students of a new, and in many ways old, integrative approach to higher education that many faculty believe will serve to effectively prepare students for work, life, and citizenship in the 21st century.

To accomplish this challenging study, the National Academies assembled a committee composed of leaders and scholars in higher education and industry with expertise in the arts, humanities, social sciences, natural sciences, engineering, and medicine—and the intersections among these disciplines—whose affiliations reflected the diversity of types of institutions in American higher education. I have learned an enormous amount from these colleagues and, now, friends, and am indebted to them for their tireless efforts, knowledge, insights, and savvy. The study was also made possible by the superb professionals from the National Academies, and the leadership of the Study Director, Ashley Bear, and the Director of the BHEW, Tom Rudin, as well as the research efforts of Irene Ngun and Kellyann Jones-Jamtgaard, and the logistical expertise of Austen Applegate.

To inform our deliberations, we heard from experts from beyond the committee, held public sessions in three cities, commissioned literature reviews, and heard from faculty across the country who submitted responses to a "Dear Colleague" letter asking for evidence and input from the broader higher education community.

WHAT DID WE FIND?

Assessing student learning outcomes across the breadth of American higher education is a daunting task, confounded by the number and types of institutions, the broadly varying backgrounds of the students matriculating, and, importantly, the fact that curricular decisions are—appropriately—in the hands of local faculty members, not subject to any broad, national consensus except in the case of accreditation of specific disciplines. For these reasons, as well as the lack of agreement on the most effective ways to assess student learning outcomes, we found that large, controlled, randomized testing of the hypothesis that integrated education would lead to educational and employment benefits are rare and likely to remain so. Nonetheless, we found abundant narrative and anecdotal evidence, some evidence from research studies, and, very importantly, a broad, national groundswell of interest in developing approaches to integrated education. Though causal evidence on the impact of integration on students is limited, it is this committee's consensus opinion that further effort be expeditiously exerted to develop and disseminate a variety of approaches to integrated education and that further research on the impact of such programs and courses on students be supported and conducted.

Ultimately, the decision will rest with the faculty of American higher education. We hope that our faculty colleagues will take the time to examine this report and will thereby join with us in further exploring the value and role of integrated education. We believe the future of our nation will be affected by our collective decisions.

David J. Skorton
Chair
Committee on Integrating Higher Education in the Arts, Humanities, Sciences, Engineering, and Medicine

Acknowledgments

The Committee on Integrating Higher Education in the Arts, Humanities, Sciences, Engineering, and Medicine would like to acknowledge and thank the many people who made this study possible. First, we would like to acknowledge the support of the standing National Academies Board on Higher Education and Workforce (BHEW), which offered oversight for the study. Secondly, we would like to acknowledge that this report was informed by the efforts of the many people who shared their data, insights, ideas, enthusiasm, and expertise with the committee. We would especially like to thank the following people (listed alphabetically) who presented at the committee's meetings and information-gathering workshops:

William "Bro" Adams, National Endowment for the Humanities
Amy Banzaert, Department of Engineering, Wellesley College
Dan Brabander, Wellesley College
Fritz Breithaupt, Germanic Studies, Indiana University Bloomington
Loren B. Byrne, Roger Williams University
Rita Charon, Program in Narrative Medicine, Columbia University
Helen Drinan, Simmons College
Ethan Eagle, Department of Mechanical Engineering, Wayne State University
Pam Eddinger, Bunker Hill Community College
David Edwards, School of Engineering and Applied Sciences at Harvard University and Le Laboratoire
Bret Eynon, LaGuardia Community College
Ed Finn, School of Arts, Media + Engineering, Arizona State University

Marie Adamson Flesher, The Ohio State University
Howard Gardner, Harvard Graduate School of Education
David Guston, Future of Innovation in Society, Arizona State University
Kevin Hamilton, College of Fine and Applied Arts at the University of Illinois, Urbana-Champaign
Maria Hesse, Academic Partnerships, Arizona State University
Ed Hundert, Harvard Medical School
Joel Katz, Internal Medicine Residency Program, Harvard Medical School
JoAnn Kuchera-Morin, Media Arts & Technology and Music, University of California
Liz Lerman, Liz Lerman Dance Exchange and Herberger Institute for Design and the Arts, Arizona State University
Richard K. Miller, President and Professor, Olin College of Engineering
Michelle Morse, Partners In Health, EqualHealth, Brigham and Women's Internal Medicine Residency
Dan Nathan-Roberts, Industrial and Systems Engineering, San José State University
Scott Page, Departments of Political Science and Economics at the University of Michigan
Lee Pelton, Emerson College
Peter Pesic, Science Institute, St. John's College
Andrea Polli, Art and Ecology, University of New Mexico
Catherine Pride, Middlesex Community College
Bob Pura, Greenfield Community College
William Ray, The Ohio State University
Robert Root-Bernstein, Michigan State University
Joaquin Ruiz, College of Letters, Arts, and Science, University of Arizona
Ben Schmidt, Northeastern University
Vandana Singh, Framingham State University
Jim Spohrer, Cognitive OpenTech, IBM Research – Almaden
Raymond Tymas-Jones, University of Utah College of Fine Arts
Rick Vaz, Center for Project-Based Learning, Worcester Polytechnic Institute
David Weaver, Professor of Physics, Estrella Mountain Community College
Rosalind Williams, Massachusetts Institute of Technology
Sha Xin Wei, School of Arts, Media and Engineering, Arizona State University
Emma Smith Zbarsky, Department of Applied Mathematics, Wentworth Institute of Technology

The committee would also like to thank students from Arizona State University, **Cecilia Chou, Matt Contursi, Tess Doezema,** and **Anna Guerrero,** for sharing their experience with the committee, as well as the respondents to the committee's "Dear Colleague" letter, for all their valuable input on integrative courses and programs.

Further, the committee would like thank the sponsors that made this study possible: the Andrew W. Mellon Foundation, the National Endowment for the Humanities, the National Endowment for the Arts, the National Academy of Sciences Scientists and Engineers for the Future Fund, and the Teagle Foundation.

We would also like to express our sincere gratitude for the generosity of the hosts of the study's two regional information gathering workshops: Le Laboratoire, Cambridge, Massachusetts, and Arizona State University, Tempe, Arizona.

The committee would like to acknowledge the work of the consultants who have contributed to the report: Dr. Matthew Mayhew, Dr. Hannah Stewart-Gambino, and Dr. Jennifer Stroud-Rossman and the report writer, Steve Olson.

ACKNOWLEDGMENT OF REVIEWERS

This Consensus Study Report was reviewed in draft form by individuals chosen for their diverse perspectives and technical expertise. The purpose of this independent review is to provide candid and critical comments that will assist the National Academies of Sciences, Engineering, and Medicine in making each published report as sound as possible and to ensure that it meets the institutional standards for quality, objectivity, evidence, and responsiveness to the study charge. The review comments and draft manuscript remain confidential to protect the integrity of the deliberative process.

We wish to thank the following individuals for their review of this report: James Barber, College of William and Mary; May Berenbaum, University of Illinois at Urbana-Champaign; Rita Charon, Columbia University; Dianne Chong, Boeing Research and Technology (Retired); Michele Cuomo, Montgomery County Community College; Jerry Jacobs, University of Pennsylvania; Leah Jamieson, Purdue University; Christine Ortiz, Massachusetts Institute of Technology; Robert Pura, Greenfield Community College; Robert Root-Bernstein, Michigan State University; Jack Schultz, University of Missouri; and James Spohrer, IBM.

Finally, we thank the staff of this project for their valuable leadership, input, and support. Specifically, we would like to thank Program Officer and Study Director, Ashley Bear; BHEW Director, Tom Rudin; Research Associate, Irene Ngun; Christine Mirzayan Science and Technology Fellow, Kellyann Jones-Jamtgaard; Senior Program Assistant, Austen Applegate; Senior Advisor, Jay Labov; and the Director of the Cultural Programs for the National Academies, J. D. Talasek.

Although the reviewers listed above provided many constructive comments and suggestions, they were not asked to endorse the conclusions or recommendations of this report nor did they see the final draft before its release. The review of this report was overseen by Maryellen Giger, University of Chicago, and Cora Marrett, University of Wisconsin-Madison. They were responsible for making certain that an independent examination of this report was carried out in accordance with the standards of the National Academies and that all review comments were carefully considered. Responsibility for the final content rests entirely with the authoring committee and the National Academies.

Contents

Boxes, Figures, and Tables

BOXES

FIGURES

TABLES

IMAGES FROM GALLERY OF ILLUMINATING AND INSPIRATIONAL INTEGRATIVE PRACTICES IN HIGHER EDUCATION

Summary

Albert Einstein once said, "all religions, arts, and sciences are branches from the same tree" (Einstein, 2006, p. 7). This holistic view of all human knowledge and inquiry as fundamentally connected is reflected in the history of higher education—from the traditions of Socrates and Aristotle, to the era of industrialization, to the present day. This view holds that a broad and interwoven education is essential to the preparation of citizens for life, work, and civic participation. An educated and open mind empowers the individual to separate truth from falsehood, superstition and bias from fact, and logic from illogic.

In the United States, broad study in an array of different disciplines—including the arts, humanities, sciences, and mathematics—as well as in-depth study within a special area of interest, has been a defining characteristic of higher education. But over time, the curriculum at many colleges and universities has become focused and fragmented along disciplinary lines. This change in higher education has been driven, in part, by increasing specialization in the academic disciplines and the associated cultural and administrative structure of modern colleges and universities. Now many leaders, faculty, scholars, and students have been asking whether higher education has moved too far from its integrative tradition toward an approach heavily rooted in disciplinary "silos." These silos represent what many see as an artificial separation of academic disciplines.

This study examined an important trend in higher education: efforts to return to—or in some cases to preserve—a more integrative model of higher education that proponents argue will better prepare students for work, life, and citizenship. This integrative model *intentionally* seeks to bridge the knowledge, modes of inquiry, and pedagogies from multiple disciplines—the humanities, arts, sciences, engineering, technology, mathematics, and medicine—within the context of a single course or program of study. In such a model, professors help students to make the connections between these disciplines in an effort to enrich and improve learning. A diverse array of colleges and universities now offer students integrative courses and programs, and many faculty are enthusiastic advocates for this educational approach. But this movement in higher education raises an important question: what is the impact of these curricular approaches on students?

To address this question, the National Academies of Sciences, Engineering, and Medicine formed a 22-member committee to examine "the evidence behind the assertion that educational programs that mutually integrate learning experiences in the humanities and arts with science, technology, engineering, mathematics, and medicine (STEMM) lead to improved educational and career outcomes for undergraduate and graduate students." The committee conducted an in-depth review and analysis of the state of knowledge on the impact of integrative approaches on students.

EVIDENCE FOR THE OUTCOMES OF INTEGRATION

The case for integrating the arts, humanities, and STEMM fields in higher education must ultimately rest on evidence that is sufficiently convincing to inspire the adoption of such models in undergraduate and graduate education. Over the course of our study, we examined a broad array of evidence related to integration to draw our conclusions, including the research literature, examples of integrative programs, input from experts who met with the committee and responded to a "Dear Colleague" letter, public input, employer surveys, and other information relevant to the effort. In the spirit of integration, we also examined diverse forms of evidence, including personal testimony from faculty, administrators, students, and employers on the value of an integrative approach to education; essays and thought pieces that make logical arguments for integration based on observations and evidence about common practices in higher education today; and formal and informal evaluations of courses and programs carried out by institutions.

Despite the many challenges of assessing the impact of integrative educational approaches on students, the available research does permit several broad conclusions to be made:

- Aggregate evidence indicates that some approaches that integrate the humanities and arts with STEM have been associated with positive learning outcomes. Among the outcomes reported are increased critical thinking abilities, higher-order thinking and deeper learning, content mastery, problem solving, teamwork and communication skills, improved visuospatial reasoning, and general engagement and enjoyment of learning (see Tables 6-1 and 6-2 for an overview of some of the learning outcomes associated with specific integrative approaches).

- The integration of STEMM content and pedagogies into the curricula of students pursuing the humanities and arts may improve science and technology literacy and can provide new tools and perspectives for artistic and humanistic scholarship and practice.
- The integration of the arts and humanities with medical training is associated with outcomes such as increased empathy, resilience, and teamwork; improved visual diagnostic skills; increased tolerance for ambiguity; and increased interest in communication skills.
- Many faculty have come to recognize the benefits of integrating arts and humanities activities with STEMM fields and offer first-hand testimony to the positive student learning outcomes they observe as associated with integrative curricula.
- Abundant interest and enthusiasm exist for integration within higher education, as evidenced by the groundswell of programs at colleges and universities in various sectors of American higher education (see "Compendium of Programs and Courses That Integrate the Humanities, Arts, and STEMM" at https://www.nap.edu/catalog/24988 for a list of 218 examples that the committee found illustrative).

An important observation was that the kinds of outcomes associated with certain integrative approaches in higher education—including written and oral communication skills, teamwork skills, ethical decision making, critical thinking, and the ability to apply knowledge in real-world settings—are the educational outcomes that many employers are asking for today. Employer surveys consistently show that employers are asking for graduates with more than deep technical expertise or familiarity with a particular technology. They are looking for well-rounded individuals with a holistic education who can take on complex problems and understand the needs, desires, and motivations of others. Interestingly, these learning goals and competences are similarly valued by institutions of higher education. A 2016 survey released by the Association of American Colleges and Universities (AAC&U) found that nearly all AAC&U member institutions—which constitute a majority of 4-year colleges and universities in the United States—have adopted a common set of learning outcomes for all their undergraduate students (Hart Research Associates, 2016). Shared learning outcomes included writing and oral communication skills, critical thinking and analytical reasoning skills, ethical reasoning skills, and "integration of learning across disciplines" (Hart Research Associates, 2016, p. 4), among others (see Figure 2-3).

Many of the observations and conclusions made of integration at the undergraduate level apply as well to graduate education. In recent years, some have argued that the traditional, disciplinary approach to graduate education may not equip students with the awareness, knowledge, and skills needed to approach, frame, and solve increasingly complicated problems. Preparing the next generation of graduate students to tackle the problems of the twenty-first century may necessitate a shift toward integration in graduate research and education. Though most graduate programs today focus on a single discipline or subdiscipline, interdisciplinary graduate programs have emerged in recent years, and many schools are working to promote greater interdisciplinarity in graduate training and scholarship. Also, established integrative fields, such as science, technology, and society; sustainability; women's studies; human–computer interaction; bioethics; and many others, offer models of successful integrative graduate-level programs. The committee observed that one important outcome of integrative graduate education is greater institutional capacity to produce future faculty members who are well prepared to provide integrative education.

Given the evidence that is currently available about the potential of integration to produce positive learning outcomes, and based in the consensus opinion of the committee members, we have drawn the following conclusions and recommendations.

SUPPORT FOR INTEGRATIVE APPROACHES

An emerging body of evidence suggests that integration of the arts, humanities, and STEMM fields in higher education is associated with positive learning outcomes that may help students enter the workforce, live enriched lives, and become active and informed members of a modern democracy. While the current evidence base limits our ability to draw causal links between integrative curricula in higher education and student learning and career outcomes, we believe it is important to acknowledge how difficult it is to carry out causal studies on educational interventions and how rarely any curriculum in higher education is evaluated. In light of these realities, and the fact that evaluation will depend on the existence of integrative programs, we do not believe it is practical for institutions with an interest in pursuing more integrative approaches to wait for more robust causal evidence before adopting, supporting, and evaluating integrative programs. In short, this committee has concluded that the available evidence is sufficient to urge support for courses and programs that integrate the arts and humanities with STEMM in higher education. Therefore, we recommend the following:

Individual campus departments and schools, campus-wide teams, and campus–employer collaborators should consider developing and implementing new models and programs that integrate the STEMM fields, the arts, and the humanities.

Institutions should work to sustain ongoing integrative efforts that have shown promise, including but not limited to, new integrative models of general education.

New designs for general education should consider incorporating interdisciplinary, multidisciplinary, and transdisciplinary integration, emphasizing applied and engaged learning and connections between general education and specialized learning throughout the undergraduate years and across the arts, humanities, and STEMM disciplines.

Institutions interested in supporting integrative curricular models should set aside resources for the hiring, research, teaching activities, and professional development of faculty who are capable of teaching integrative courses or programs.

Both federal and private funders should recognize the significant role they can and do play in driving integrative teaching, learning, and research. We urge funders to take leadership in supporting integration by prioritizing and dedicating funding for novel, experimental, and expanded efforts to integrate the arts, humanities, and STEMM disciplines and the evaluation of such efforts. Sustained support will be necessary to understand the long-term impact of integrative approaches.

EVALUATING INTEGRATIVE COURSES AND PROGRAMS

Given the limited, but promising, evidence for positive learning and career outcomes associated with integration in higher education, the committee is urging that a new nationwide effort be undertaken to collect a robust and multifaceted body of evidence that the broader educational community can accept, embrace, and apply to specific settings throughout the huge and complex landscape of American higher education. We recommend the following:

Those interested in fostering disciplinary integration in higher education, including faculty and administrators, should work with scholars of higher education and experts in the humanities, arts, and STEMM fields to establish agreement on the expected learning outcomes of an integrative educational experience and work to design approaches to assessment.

Stakeholders (e.g., faculty, administrators, and scholars of higher education research) should employ multiple forms of inquiry and evaluation when assessing courses and programs that integrate the humanities, arts, and STEMM fields, including qualitative, quantitative, narrative, expert opinion, and portfolio-based evidence. Stakeholders should also consider developing new evaluation methodologies for integrative courses and programs.

Given the challenges of conducting controlled, randomized, longitudinal research on integrative higher education programs, we recommend two potential ways forward: (1) institutions with specific expertise in student learning outcomes (e.g., schools of education) could take a leadership role in future research endeavors, and (2) several institutions could form a multisite collaboration under the auspices of a national organization (e.g., a higher education association) to carry out a coordinated research effort. In either case, efforts to identify the appropriate expertise and support necessary to conduct such research should be a priority.

Institutions should perform a cultural audit of courses, programs, and spaces on campus where integration is already taking place, partnering with student affairs professionals to evaluate programs and initiatives intended to integrate learning between the class and nonclassroom environment, and working with teaching and learning centers to develop curricula for faculty charged with teaching for or within an integrative experience.

Institutions and employers should collaborate to better understand how graduates who participated in courses and programs that integrate the humanities, arts, and STEMM fields fare in the workplace throughout their careers. We recommend the following four areas for such collaboration:

1. Institutions should survey alumni to gain a sense of how their education, particularly the integrative aspects of their programs, has served them in work, life, and civic engagement. Institutions should share the results of such surveys with employers.

2. Employers should gather and share with institutions information about the educational experiences, especially integrative experiences, that lead to employee success.

3. Where possible, institutions and employers should find ways to collaborate on these activities.

4. Professional artistic, humanistic, scientific, and engineering societies should work together to build, document, and study integrative

pilot programs and models to support student learning and innovative scholarship at the intersection of disciplines.

ENHANCING INCLUSIVITY THROUGH INTEGRATIVE COURSES AND PROGRAMS

As our committee considered the evidence associated with integrative learning in the arts, humanities, and STEMM fields, we also sought evidence of the benefits of integrative learning to groups of people who have been historically underserved by higher education. Women, people with disabilities, and population groups including African Americans, Latinos, and indigenous people have not participated equitably in some areas of the arts, humanities, and STEMM fields. Any new movement in higher education, we believe, must ensure that it prepares all students to prosper economically, contribute civically, and flourish personally.

Issues of equity and diversity in higher education intersect with the goals of disciplinary integration. One of the goals of disciplinary integration is to make connections between STEMM fields and other disciplines so that STEMM subjects (and hopefully STEMM careers) become more appealing to groups traditionally underrepresented, and at times actively excluded, from STEMM fields. In our analysis of the evidence on the impact of integrative educational programs, the committee found several instances in which the integration of the arts and humanities with STEM was associated with particular benefits for women and underrepresented minorities. Therefore, we recommend the following:

Further research should focus on how integrative educational models can promote the representation of women and underrepresented minorities in specific areas of STEMM fields, the arts, and the humanities, and all research efforts should account for whether the benefits of an integrative approach are realized equitably.

REMOVING THE BARRIERS TO INTEGRATIVE APPROACHES

Certain internal and external pressures are placed on universities that are likely to drive disciplinary segregation and serve as barriers to integration. To facilitate integration and overcome these barriers, we recommend the following:

When implementing integrative curricula, faculty, administrators, and accrediting bodies need to explore, identify, and mitigate constraints (e.g., tenure and promotion criteria, institutional budget models, workloads,

accreditation, and funding sources) that hinder integrative efforts in higher education.

Academic thought leaders working to facilitate integrative curricular models should initiate conversations with the key accrediting organizations for STEMM, the arts, and higher education to ensure that the disciplinary structures and mandates imposed by the accreditation process do not thwart efforts to move toward more integrative program offerings.

Professional development of current and future faculty is needed to promote integrated learning, given the additional complexity of pedagogy in integrated courses and programs, and research on effective pedagogical practices for integrated learning should be expanded.

TOWARD A MORE INTEGRATED FUTURE

The fragmentation of knowledge and learning was a historical process, and the future can depart from that past. Given that today's challenges and opportunities are at once technical and human, addressing them calls for the full range of human knowledge and creativity. Future professionals and citizens need to see when specialized approaches are valuable and when they are limiting, find synergies at the intersections between diverse fields, create and communicate novel solutions, and empathize with the experiences of others. The committee views the integration of the arts, humanities, and STEMM fields in higher education as a promising avenue to help create this future.

1

Challenges and Opportunities for Integrating Established Disciplines in Higher Education

Albert Einstein once said, "all religions, arts, and sciences are branches from the same tree."[1] (Einstein, 2006). This holistic view of all human knowledge and inquiry as fundamentally connected is reflected in the history of higher education—from the traditions of Socrates and Aristotle, to the era of industrialization, to the present day. This view holds that a broad and interwoven education is essential to the preparation of citizens for life, work, and civic participation. An educated and open mind empowers the individual to separate truth from falsehood, superstition and bias from fact, and logic from illogic.

In the United States, broad study in an array of different disciplines—including the arts, humanities, science, and mathematics—as well as in-depth study within a special area of interest, has been a defining characteristic of higher education. But over time, the curriculum at many colleges and universities has become focused and fragmented along disciplinary lines in such a way that some faculty have begun to ask whether students are now struggling to see the connections between different forms of knowledge and approaches to human inquiry. This change in higher education has been driven, in part, by increasing specialization in the academic disciplines and the associated cultural and administrative structure of modern colleges and universities.

---

[1] Einstein's statement about the branches of the tree was made in a letter to the YMCA in October 1937 against a backdrop of growing fascist power in central Europe. Einstein warned of the dangerous implications of living in a society where long-established foundations of knowledge were corrupted, manipulated, and coerced by political forces.

There is little doubt that disciplinary specialization has helped produce many of the achievements of the past century. Researchers in all academic disciplines have been able to delve more deeply into their areas of expertise, grappling with ever more specialized and fundamental problems. Today, the academic disciplines continue to be extraordinary wellsprings of expertise, creativity, and innovation and have provided a critical infrastructure for educating a highly skilled, if increasingly specialized, workforce. Since World War II, the research and training done at colleges and universities have come to be seen as crucial to economic competitiveness, national security, and social well-being.

Yet many leaders, faculty, scholars, and students have been asking in recent years whether higher education has moved too far from its integrative tradition toward an approach heavily rooted in disciplinary "silos." These silos represent what many see as an artificial separation of academic disciplines.[2] More than 50 years ago, University of California President Clark Kerr sounded an alarm about the disciplinary segregation of the university. The university had become a "multiversity," he said, held together more by a unitary administrative structure and budget than by a collective commitment to truth or to a notion that knowledge is essentially integrated. As he put it, the modern research university consists of "a series of individual faculty entrepreneurs held together by a common grievance over parking"(Kerr, 2001, p. 15). Indeed, the committee found that the administrative and budgetary structures at many institutions lend themselves to the disciplinary segregation of knowledge and learning. Such observations suggest that the depth and breadth of students' exposure to other disciplinary traditions—and, in turn, other ways of understanding and questioning their world—may be limited at many schools.

This report examines an important trend in higher education: efforts to return to—or in some cases to preserve—a more integrative model of higher education that proponents argue will better prepare students for work, life, and citizenship. The model of integrative education examined in this report is both new and old. It is old in that it is rooted in a long-standing tradition of integration in education. It is new in the way it *intentionally* seeks to integrate knowledge to meet the challenges and opportunities of the twenty-first century. This movement goes beyond the general education curriculum found at almost every institution of higher education in the United States, in which students take several disconnected courses in different disciplines outside their major. In this new model, the knowledge, modes of inquiry,

[2] Bass and Eynon (2017) have used this term in a narrower but similar context, writing, "The most influential commercial applications of educational technology have largely been disintegrative—i.e., modular, focusing mainly on efficiency and productivity, and addressing narrow dimensions of learning" (p. 3).

and pedagogies from multiple disciplines are brought together within the context of a single course or program of study. In such a model, professors help students to make the connections between these disciplines in an effort to enrich and improve learning. Some call such efforts "STEAM"[3] or "SHTEAM"—acronyms that variously combine science, technology, engineering, the arts, mathematics, and the humanities. Others use the older term "liberal education." This committee refers to the integration of knowledge and pedagogies in the humanities, arts, social sciences, natural sciences, engineering, technology, mathematics, and medicine[4] simply as "integration."

The committee's task was to examine the impact of integrative educational experiences on student learning and career outcomes. We focused our examination of integration on the evidence related to the nature and impact of integrative courses and programs as they relate to the learning outcomes and competencies currently being called for by both employers and institutions of higher education. These include learning outcomes such as written and oral communication skills, teamwork skills, ethical decision making, critical thinking, and the ability to apply knowledge in real-world settings. Though we offer a discussion of the multiple rationales offered by proponents for the value of an integrative approach, and ultimately conclude that integration is one model that shows promise for meeting the broad educational goals shared by institutions of higher education and employers, we do not argue in this report that the disciplines are not of value or that integrative models should necessarily supplant discipline-based courses and programs. Indeed, the committee was not charged with examining other educational approaches in detail. While it is possible, and perhaps even likely, that other educational models may yield similar positive student learning outcomes to those associated with integration, this was not the focus of this study.

Proponents of integration offer multiple rationales for the value of an integrative approach in higher education. Some argue that integration in higher education is needed to address the unprecedented global challenges

[3] STEAM is a movement toward greater integration of the arts and humanities with STEM subjects that often takes place within the context of K–12 education. Though this study is focused on higher education, so that integration at the K–12 level was out of the scope of the study, the committee acknowledges that integration at the K–12 level is worthy of study and that K–12 education and higher education are fundamentally connected within the U.S. education system.

[4] Although medicine is a postgraduate professional degree such as law or journalism, the committee believes that medicine fits appropriately within the scope of this study because of the groundswell of programs that integrate the arts and humanities with medical training and because of the nature of medicine as an applied, humanistic STEM discipline, much like engineering.

and opportunities of our time (United Nations Development Programme, 2016). Some insist that an integrative approach will better prepare graduates for employment or that it will better prepare graduates for engaged citizenry. Others simply observe that an integrative approach makes learning more fun, engaging, and relevant to students. The committee views these different rationales as deeply connected and in no way mutually exclusive.

Those who see integration as necessary for addressing the challenges of our time argue that the transformative changes of the past and the grand challenges of the present cut across multiple dimensions of human life—material, economic, environmental, social, cultural, technical, political, medical, aesthetic, and moral. They argue that to address the challenges and seize the opportunities of our time will require an education that draws upon all forms of human knowledge creation—the artistic, humanistic, scientific, technological, and medical—and the intersections and connections among them. The world is a human world, and scientific expertise in isolation offers an essential but incomplete foundation for guiding humanity's future (Box 1-1). At the same time, humanistic and artistic engagement with the world that neglects its significant and ever-expanding scientific and technological dimensions likewise offers an incomplete picture. For example, modern information and communication technologies, developed and scaled through partnerships between engineers, scientists, artists, and humanists, have altered the ways people work, access goods and services, relate to one another, and participate in public life. The influences of these technologies on our society demand more reasoned and inclusive forms of civic engagement, even as they have facilitated new forms of division, distrust, and extremism. Similarly, advances in genome editing are lending new urgency and complexity to age-old questions of human rights, integrity, and dignity. Or take the example of climate change. As rapid, reliable, and inexpensive access to energy has become embedded in the rhythms and norms of social life, the transition to a less carbon-dependent economy is as much a cultural and moral challenge as it is a technological one (Box 1-2). Thus, proponents of educational integration contend that an integration of knowledge is necessary to address the challenges of our time. They argue that narrowed conceptions of the nature of problems—the hallmark of disciplinary specialization—may produce a range of possible solutions that are similarly constrained. As physicist and Nobel laureate Murray Gell-Mann has suggested, "we must rid ourselves of the notion that careful study of a problem based on a narrow range of issues is the only kind of work to be taken seriously, while integrative thinking is to be relegated to cocktail party conversation" (Schellnhuber, 2010, p. 3).

The argument that an integrative approach to education equips students with the knowledge, skills, and competencies to deal with the complex, multidimensional challenges of the world outside of campus also applies to

BOX 1-1
The Power of STEMM, the Arts, and Humanities in Responding to the Ebola Epidemic

When an epidemic caused by the Ebola virus was ravaging western Africa in 2014–2016, the STEMM disciplines (including the social sciences) were critical in responding to the disease. But the arts and humanities also played major roles in ending the epidemic. Popular songs broadcast widely on radio stations taught people how to keep from getting infected. Interdisciplinary teams designed protective suits for health care workers that were easier to use. Los Angeles-based professor and artist Mary Beth Heffernan flew to Liberia with cameras and printers and began attaching photographs of health care workers to the fronts of their protective suits so that people would be less alarmed by the impersonal, white-clad medics (see the video at http://www.letitripple.org/films/adaptable-mind/). The importance of the arts and humanities in responding to the Ebola epidemic reflects a much more general observation. The arts and humanities are essential partners with STEMM fields in solving problems that occur in all sectors of society and at all levels, from the personal to the global.

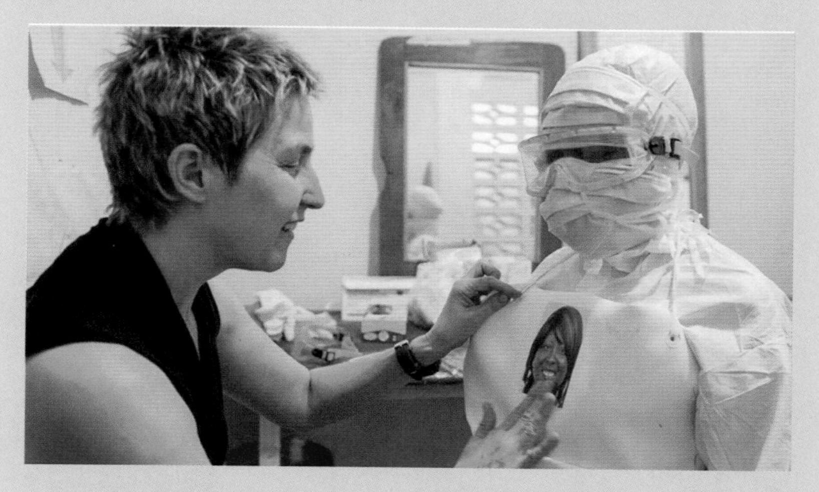

Artist Mary Beth Heffernan affixes PPE Portrait to Zoe Dewalt, RN, in the donning room of ELWA 2 Ebola Treatment Unit, March 12, 2015, Paynesville, Liberia.

SOURCE: Mary Beth Heffernan, PPE Portrait Project, 2014-2015.
Photo credit: Marc Campos.

BOX 1-2
The Need for Disciplinary Integration in Addressing Climate Change

Climate change provides a specific example of the ways in which today's challenges and opportunities transgress the boundaries between disciplines. Concentrations of carbon dioxide and other greenhouse gases in the atmosphere are interwoven with social inequality (Islam and Winkel, 2016). Present levels reflect different rates of energy-intensive industrialization and the unequal global distribution of its benefits (Economist, 2017). At the same time, current emissions reflect cultural and social ideas, such as what it means to be a developed economy and what relationships and responsibilities members of a society have to each other, to future generations, to the global community, and to the planet. Models and measures of climatic effects capture only one dimension of this complex human challenge. Addressing climate change will require contending with the meanings of cultural habits and histories and seeking new ways for national and global communities to understand, reimagine, and give expression to visions of responsibility for the future. In short, the challenges and opportunities of addressing climate change are integrated. The "smog meringue" was developed to allow the public to taste the impact of climate change. This effort is led by CoClimate, an artist-led think tank and strategic design studio whose mission is to "build essential questions about climate change." See the smog tasting project at: http://genomicgastronomy.com/work/2017-2/smog-tasting-take-out/.

SOURCE: The Center for Genomic Gastronomy.

the world of work. Importantly, some of the most vocal calls for reform in higher education have come from industry, especially from the technology, engineering, and business sectors. As described in Chapter 2 of this report, employers have noted a mismatch between the skills they want in their employees and the skills many graduates leave higher education with today. Employers insist that the acquisition of highly specialized, discipline-specific skills should not replace equally important and more lasting sets of skills, such as writing and communication, critical thinking, the ability to work on teams, ethical and cultural awareness, and lifelong learning attitudes. Chapter 5 of this report illustrates that such outcomes are associated with courses and programs in higher education that intentionally integrate the arts and humanities with science, technology, engineering, mathematics, and medicine.

Beyond employment, many argue that an integrative approach to higher education will better prepare students to be educated, informed, and active citizens of a modern democracy. To make informed decisions, citizens need to be able to distinguish truth from falsehood and right from wrong. Citizens need to appreciate some of the basic methods and conclusions of science and technology as well as the social, cultural, political, aesthetic, and ethical implications of scientific and technological advances. They need to be able to make sense of other points of view and rapidly changing circumstances. They need to understand the issues addressed through public policies and the impact of decisions in multiple dimensions. Proponents of integration argue that citizens with a broad education that includes the arts, humanities, science, technology, engineering, mathematics, and medicine will be better equipped to make decisions on complex societal issues. For example, the political theorist Danielle Allen has argued that people should explore political questions by trying to make the best possible argument, on any given question, from the perspective of someone with whom they disagree or whose experience of life differs fundamentally from their own (Allen, 2004). She also asks us to ask ourselves, when we interact with strangers, whether we have treated them as we would a friend. Advocates of integration see broad, interdisciplinary experiences in college as a route to mastering these difficult skills.

The committee also considered the connection between an integrative approach to education and the importance of diversity and inclusion in higher education. Conceptual arguments for the educational benefits of diversity overlap with the arguments for integrative approaches to learning. Each is rooted in the idea that higher education is an important place to learn about oneself and others by grappling with conflicting arguments, evidence, and experiences, and by gaining an understanding of different viewpoints and perspectives. Disciplinary integration brings together diverse viewpoints, epistemologies, traditions, and pedagogies so that students gain a more holistic understanding of a topic. In addition, by expanding the pedagogical repertoire of science and engineering courses and programs, educational integration might help to improve the participation of underrepresented groups in the sciences and engineering, such as women and certain racial and ethnic minorities. Traditional science and engineering pedagogies have been shown to discourage women and people of color from pursuing science and engineering majors and careers (PCAST, 2012). The hope is that combinations of pedagogies and subject matter will promote diversity and inclusion. The committee explores this argument and the evidence to support it throughout the report.

Inspired by the multiple arguments in favor of a more integrative model of higher education, many educators and administrators have responded by offering students courses and programs that intentionally integrate knowledge from the arts and humanities with the natural sciences, social sciences, technology, engineering, mathematics, and medical disciplines (see "Compendium of Programs and Courses That Integrate the Humanities, Arts, and STEMM" available at https://www.nap.edu/catalog/24988 under the Resources tab for a list of 218 examples that the committee found illustrative). Such courses and programs are diverse. They take inter-, multi-, and transdisciplinary approaches to integrate various disciplines, pedagogies, and curricula with the goal of promoting positive student outcomes. Some integrative courses and programs are relatively new, while others have been offered to students for decades. Indeed, many integrative efforts have led to the establishment of "interdisciplines" such as Science, Technology, and Society; Gender Studies; Bioethics; and Computer–Human Interaction, among many others. Such mature fields of inter- and transdisciplinary scholarship have historically arisen at the intersections of existing fields. These interdisciplines represent the potential for academic innovation through integration.

This study reflects a growing concern that an approach to higher education that favors disciplinary segregation is poorly suited to the challenges and opportunities of our time. It examines an alternative approach, which proponents argue may increase long-term opportunities for individual attainment and societal mobility. This report explores the wealth of innovative and promising efforts in undergraduate and graduate education that bridge divisions between these disciplines—what we refer to simply as "integration." The study is motivated by the recognition that the demands of the day require that colleges and universities revisit, reflect upon, and potentially dominant educational approaches in American institutions of higher learning to ensure that students are graduating with the knowledge, skills, and competencies they will need to thrive in work, life, and civic participation.

Einstein's statement about the branches of the tree was made in a letter to the YMCA in October 1937 against a backdrop of growing fascist power in central Europe. Einstein warned of the dangerous implications of living in a society where long-established foundations of knowledge were corrupted, manipulated, and coerced by political forces. The discussions of this study committee from July 2016 to October 2017 about the integration of science, technology, engineering, mathematics, and medicine (STEMM) fields, the arts, and the humanities occurred against a backdrop of public challenges of long-established conventions about factual information, the use of evidence, and consensus knowledge. For Einstein, the proper response to threats to knowledge was to return to nineteenth-century principles that would allow societies "to preserve right and the dignity of man" and "allow us to rejoice in humanity" (Einstein, 1956, p. 8). Our committee has explored a different aspect of our common humanity—the integrated nature of human inquiry and its potential to prepare graduates from higher education for the challenges and opportunities of our time.

In this report, we delve more deeply into the rationale for an integrative approach to higher education and examine the evidence behind the assertion that such programs and courses lead to improved educational and career outcomes for students. This process gave the committee a deeper appreciation for the value of different forms of evidence and the practical realities of evidence-based decision making. After an intensive examination of the available evidence, and after deep discussions about the nature of evidence and the limits of measurement, we concluded that the existing evidence on integrative courses and programs points to very encouraging student learning outcomes that align with the twenty first–century professional skills that employers are seeking and the shared learning goals of many higher education institutions for their graduates. Further, we found abundant evidence of tremendous enthusiasm and activity among many in the higher education community to support and experiment with more integrative educational models.

THE CHARGE TO THE COMMITTEE

The increasing interest in educational integration of the humanities, arts, and STEMM disciplines among educators, students, employers, and policy makers led the National Academies of Sciences, Engineering, and Medicine in 2016 to create the Committee on Integration of Education in the Sciences, Engineering, and Medicine with the Arts and Humanities at the Undergraduate and Graduate Levels. The 22 members of our committee include representatives from academia, business, government, and nonprofit organizations with scholarly expertise in the arts, humanities, natural sciences, social sciences, engineering, and medicine, as well as expertise in the integration of these disciplines. We were charged with examining "the evidence behind the assertion that educational programs that mutually integrate learning experiences in the humanities and arts with science,[5] technology, engineering, mathematics, and medicine (STEMM) lead to improved educational and career outcomes for undergraduate and graduate students." In particular, the statement of task charged the committee to examine the following[6]:

- Evidence regarding the value of integrating more STEMM curricula and labs into the academic programs of students majoring in the humanities and arts in order to understand the following: (1) how STEMM experiences provide important knowledge about the scientific understanding of the natural world and the characteristics of new technologies, knowledge that is essential for all citizens of a modern democracy; (2) how technology contributes essentially to sound decision making across all professional fields; and (3) how STEMM experiences develop the skills of scientific thinking (a type of critical thinking), innovation, and creativity that may complement and enrich the critical thinking and creativity skills developed by the arts and humanities.
- Evidence regarding the value of integrating curricula and experiences in the arts and humanities—including , history, literature, philosophy, culture, and religion—into college and university STEMM education programs, in order to understand whether and how these experiences (1) prepare STEMM students and workers to be more effective communicators, critical thinkers, problem solvers, and leaders; (2) prepare STEMM graduates to be more creative and effective scientists, engineers, technologists, and health care providers, particularly with respect to understanding the broad social and cultural impacts of applying knowledge to address challenges and opportunities in the workplace and in their communities; and (3) develop skills of critical thinking, innovation, and creativity that may complement and enrich the skills developed by STEMM fields.
- New models and good practices for mutual integration of the arts and humanities and STEMM fields at 2-year colleges, 4-year colleges, and graduate programs, drawing heavily on an analysis of programs that have been implemented at institutions of higher education.

[5] This committee includes the social and behavioral sciences in the definition of science. See Chapter 3.

[6] Appendix II contains the complete statement of task for our charge.

Integration of the Arts and Humanities into STEMM Fields

Our charge revolves around three basic questions. The first asks whether integrating curricula and experiences in the arts and humanities—including history, literature, philosophy, culture, and religion—into college and university STEMM education programs (1) can prepare STEMM students and workers to be more effective communicators, critical thinkers, problem-solvers, and leaders; (2) can prepare STEMM graduates to be more creative and effective scientists, engineers, technologists, and health care providers, particularly with respect to understanding the broad social and cultural impacts of applying knowledge to address challenges and opportunities in the workplace and in their communities; and (3) can help students develop skills of critical thinking, innovation, and creativity that may complement and enrich the skills developed by STEMM fields. Imbedded in these questions are several hypotheses about the value of educational integration. The first and third questions relate to the idea that the integration of the humanities and arts into the curricula of students majoring in STEMM subjects will promote the development of twenty first–century skills and competencies that employers and educators are both calling for today. The second question relates to the reasoning, that when scientists, technicians, engineers, mathematicians, and health professionals understand more about human history and culture, they can draw from a deeper pool of knowledge in understanding the context of their work and in solving problems. They can avoid professional narrowness both within and outside their fields; for example, they can explain to people outside of STEMM fields what they do and why it is important. They can better understand the social origins and context of their own scientific and technological disciplines. They can learn to think broadly as well as deeply, to be curious, to contend with ambiguity, and to identify gaps and limitations in their knowledge. They can become more adept at self-expression and empathy. They can be better learners, more flexible and valuable professionals, and more enlightened citizens by virtue of having a larger social and moral context in which to do scientific and technical work.

The argument is that the arts and humanities provide instrumental benefits, such as better communication with professionals in fields outside of STEMM, although integration may provide such benefits. Rather, the contention is that success in STEMM fields is enhanced through engagement with the ideas, methods, and contributions of the arts and humanities.

Integration of STEMM Subjects into the Arts and Humanities

The second major question presented by our statement of task relates to the impact of integrating more STEMM curricula and labs into the academic programs of students majoring in the humanities and arts. Specifically, we were charged with examining whether the integration of knowledge and procedures drawn from STEMM fields into the arts and humanities promotes improved scientific and technical competency and offers new insights into the natural, physical, and human-designed worlds; an understanding of and engagement with the place of science and technology in twenty first–century life, including in its social, historical, aesthetic, and ethical dimensions; and twenty first–century skills such as teamwork, critical thinking, and problem solving (Gurnon et al., 2013; Ifenthaler et al., 2015; Jarvinen and Jarvinen, 2012; Malavé and Watson, 2000; Olds and Miller, 2004; Pollack and Korol, 2013; Stolk and Martello, 2015; Thigpen et al., 2004; Willson et al., 1995). This argument holds that integrating STEMM knowledge, concepts, and forms of critical thinking into arts and humanities disciplines could strengthen and deepen learning in much the same way that arts and humanities integration benefits learning in the STEMM disciplines. All citizens, one could argue, require a basic working understanding of the scientific and technical, as well as the cultural, details of their lives and environments and the important relationships among them. A well-educated person must have an understanding of the laws of physics, the structure of science, the ways of thinking that govern scientific and technological innovation, and the historical, social, economic, and political significance of science and technology in modern life. Artists and humanists must also understand the scientific and technological context of their craft. An understanding of science and technology can breathe relevance into the work of artists and humanists and support and inspire creative engagement in life, work, and civic engagement.

Moreover, in a century in which jobs, life, and citizenry are increasingly influenced by advances in science and technology, there are greater demands placed on *everyone* to be scientifically and technologically literate.[7] Polls demonstrate that disturbing percentages of Americans have a superficial understanding of such issues as climate change, medical research, gene

[7] There is also a growing arts and humanities "illiteracy" in this nation that may disproportionately affect students majoring in STEM subjects in higher education. Many states, in response to severe budgetary constraints and economic factors over the past several decades, cut their K–6 elementary applied arts programs altogether. California's Proposition 13, enacted in 1978, limited property taxes to 1 percent of the property's value, which shrank or eliminated many state-funded programs, including arts in education. The State of Utah removed all public elementary arts education in the late 1970s, though in the past several years it has begun to reinstate arts education at the elementary level. It should be noted that this is an integrative model, now funded at more than $10 million per year by the Utah state legislature, for integrative elementary arts education run by the Beverley Taylor Sorenson Arts Learning Program, administered through the Utah state board of education. The result of a dearth of arts instruction in primary education for four decades is that multiple generations of citizens have grown up without exposure to, or active participation in, the arts through public education.

mapping, or other complex issues in the modern world. The perceived crisis of "scientific illiteracy" among those who will fill the ranks of the citizenry—for example, teachers, parents, employers, policy makers, and voters—receives attention among STEMM educators who fear that the U.S. political structure will not be able to cope with the scientific and technological choices that are necessary in the twenty-first century. A recent report of the National Academies Board on Science Education found that, on controversial issues that require some scientific and medical understanding, such as vaccines or genetic engineering, "people tend to interpret any new information in a way that fits with their worldviews" (National Academies of Sciences, 2016, p. 94).

The need for scientific and technological literacy extends beyond just understanding issues or making informed decisions. Today in the United States technology is integrated into daily life to the degree that it shapes the very decisions we make and the options we choose to consider. For example, the examination of risks present in a voting system will not only involve knowledge about the hardware and software in use, but also require careful consideration of the limitations in a given visual display and how that might shape or influence a voter's ultimate choice of candidate given fundamental cognitive heuristics and biases. According to its advocates, this kind of capability to grasp knowledge across multiple, sometimes unconnected, domains is one of the biggest benefits to an integrative approach.

Again, the argument for the integration of STEMM into arts and humanities curricula is not just instrumental. The contention is that an agile intellectual curiosity fed and fueled by an education that positions students to be able to understand their world and participate well in public life must *include* a significant encounter with STEMM knowledge, including a sense of the place of science and technology in modern life. By this reasoning, humanities and arts students need exposure to the ideas and methods associated with STEMM fields to do their jobs and live their lives more fully.

A Mutually Supportive Relationship

From a more general perspective, proponents of integration believe that learning in STEMM fields and in the arts and humanities is mutually supportive. According to this view, STEMM fields, the arts, and the humanities contribute not only to the strength of the nation but also to the strength of each other. In part, this reflects their deep similarities. As Marilyn Deegan has written,

> The humanities have more rigour and method than they are often given credit for, and a scientist needs the kind of imagination and flair more often associated with the arts. . . . Researchers working on the human ge-

> nome, the poems of John Keats, dark matter, the Tractatus of Wittgenstein, the Bible and the movement of refugees are all engaged in the same ultra-human tasks—how do we interpret ourselves, our bodies, our minds, our environment, our history and our morality. (Deegan, 2014, p. 26)

As a specific example, students who apply their scientific understanding of light to create artistic products could also use their visual aids as a way to communicate scientific ideas and findings, creating an artistic–scientific loop that contributes to creative thinking and the ability to deal with complex problems innovatively (Shen et al., 2015).

From this broader perspective, the benefits of integration are both instrumental and intrinsic. Integrating knowledge from both sides of the "two cultures" divide (as first articulated by C. P. Snow in "The Two Cultures" (1959) and discussed in more detail in Chapter 2) helps students be more effective in using their own discipline's learning. Scientists need the skills, insights, and methods of humanists and artists to help them understand the broader implications and impacts of their work from a historical to a contemporary sociocultural perspective. Humanists and artists need the skills, insights, and methods of scientists and mathematicians to integrate alternative interpretation and communication methods about phenomena they observe, explore, and integrate into their research and creative forms of expression. From an instrumental perspective, scientists need humanists and artists to help them creatively communicate and interpret their findings and share their processes with nonscientists, while humanists and artists need scientists and mathematicians to help them compile, interpret, and incorporate data and better express themselves through their artistic mediums. Opportunities for encounter, cross-fertilization, and integration between such fields can be generative for all involved, and often in ways that are difficult to imagine and predict from within the disciplines.

These benefits, though significant, are not surprising: the work of individual scientists, engineers, and physicians is driven by their experiences as humans in the world and in society. The questions that scientists ask and seek to answer are not divorced from the human condition, history, culture, society, and aesthetics. Similarly, the work of humanists and artists is fundamentally influenced and informed by scientific and technological developments and discoveries that affect the human condition, the environment in which we live, and the course of human history. Each discipline offers scholars new tools, media, and ideas for humanistic, scientific, technological, medical, and artistic exploration and discovery. Their integration creates opportunities for innovation, greater reflexivity, and deeper understanding of the disciplines and the world those disciplines seek to explore.

THE LIMITS OF ACRONYMS

Concerns about the fragmentation and specialization of knowledge and the need for more integrative curricular approaches are not new. For instance, the acronym "STEM" (science, technology, engineering, and mathematics), which has become widely used in recent decades, itself reflects a growing movement, especially at the K–12 level, to teach science, technology, engineering, and mathematics in an integrated fashion. Such an approach can make learning more connected and relevant for students (NAE and NRC, 2014). Yet some have objected to the way STEM integrates some fields while leaving others out. For example, a common observation is that STEM should be revised to STEMM by adding medicine and other fields of study related to health, as reflected in the National Science Foundation (NSF) 2017 charge to our committee. Furthermore, while historically the NSF has included social sciences under the heading of the sciences, STEM is generally considered to include the natural sciences and more quantitative social sciences and not the qualitative and more humanistically oriented social sciences.

Proposals to integrate the arts and humanities into STEM education have given rise to the acronym "STEAM," in which the "A" indicates some form of integration with the arts. STEAM education is "largely a K–12 initiative conceived to bridge the interdisciplinarity, creativity, and innovation found in both art and science" (Lewis, 2015, p. 262). John Maeda, the former Rhode Island School of Design president who has championed the shift from STEM to STEAM, strongly believes that STEM subjects "alone will not lead to the kind of breathtaking innovation the 21st century demands" (Maeda, 2013, p. 1). The STEM to STEAM transition has the goal of fostering the "true innovation that comes with combining the mind of a scientist or technologist with that of an artist or designer" (Sousa and Pilecki, 2013, para. 2). Some interpretations of "STEAM" use the "A" to imply integration with both the arts and humanities; other acronyms, such as "SHTEAM," incorporate the humanities explicitly.

In the view of our committee, the question of which acronym captures the essential disciplinary ingredients for a good education is somewhat misleading. Focusing on which disciplines are counted in (or out) has the potential to distract from the question of what educational aims should inform curricula. Rather than taking the separation of disciplines as a given and beginning with the question of which disciplines belong at the table, our committee focused instead on forms of curricular integration that seek to transcend the limitations of disciplinary boundaries by bridging among and integrating scientific, social scientific, engineering, technological, mathematical, medical, humanistic, and artistic approaches. We focused first on the aspirations and achievements of particular efforts at integration, asking what outcomes they achieve, what novel contributions they offer to higher education, and where they fall short. Only then did we explore the ways in which disciplinary and organizational structures allow or inhibit beneficial approaches.

THE ORGANIZATION OF THE REPORT

Following this brief introduction to our task, Chapter 2 further explores the many rationales for an integrative approach to higher education in light of the widely held goals of higher education today. The chapter examines the history and purposes of higher education and considers the demands placed on higher education by employers and students, as well as how the needs and desires of employers and students align with the goals of institutions of higher education. It also explores the innovation-based arguments for an integrative approach in higher education and how integration of the disciplines relates to issues of equity and diversity in education.

In Chapter 3, the committee offers definitions of the humanities, arts, science, engineering, and medicine and describes the characteristics of integration.

Chapter 4 discusses the meaning and nature of "evidence," the value of considering multiple forms of evidence, and the challenges of collecting evidence in real-world contexts and of generalizing the "evidence of improved educational and career outcomes" that is articulated in our Statement of Task when different stakeholders have different interpretations of positive educational outcomes, measure outcomes in different ways, approach integrative teaching and learning using different pedagogical structures, and integrate different disciplines in a multitude of ways.

Chapter 5 offers an overview of the many cultural and administrative barriers to integration in higher education and discusses strategies for overcoming such barriers.

In Chapter 6, the committee reviews the existing evidence on the impact of integrative educational approaches in undergraduate education, while Chapter 7 focuses on the impact of integrative graduate and medical education on students. Chapter 8 presents our conclusions drawn from the evidence we have reviewed and our consensus recommendations informed by this evidence.

We also include in the report a "Gallery of Illuminating and Inspirational Integrative Practices in Higher Education," which offers images and descriptions of artistic and humanistic scholarship, education, and practice that have been inspired, influenced, or supported by STEM knowledge, processes, and tools.

2

Higher Education and the Demands of the Twenty-First Century[1]

Throughout our country's history there has been a healthy tension between an education focused on the development of an enlightened and engaged citizenry and a more specialized, practical education for workforce development. In recent years, much of the public discourse in the United States has focused on the role of higher education in preparing students to enter the workforce. Indeed, the surveys reviewed in this chapter show that many Americans view higher education as a path to a "good job." Yet the evidence reviewed in this chapter also suggests that the educational outcomes employers are asking for today—including written and oral communication skills, teamwork skills, ethical decision making, critical thinking, and the ability to apply knowledge in real-world settings—are the same kinds of learning outcomes that many institutions of higher education believe will prepare graduates for work, life, and active, engaged citizenship. As Chapter 6 of this report will demonstrate, these are also the kinds of learning outcomes associated with certain integrative approaches in higher education.

This chapter provides a brief history of higher education in the United States and considers the relationship between disciplinary integration and issues of equity and diversity in higher education. It describes administrative structures and pressures in higher education that can drive the establishment of disciplinary "silos" and looks at what students and employers say they want from higher education. It concludes by considering some of the

[1] The committee would like to thank research consultants Hannah Stewart-Gambino and Jenn Stroud Rossmann for their significant contributions to this chapter.

ways in which higher education is changing to respond to the challenges and opportunities of our time. This discussion frames the other chapters in this report by articulating the kinds of learning outcomes that students need to be successful in their lives and careers today.

A BRIEF HISTORICAL OVERVIEW OF HIGHER EDUCATION IN THE UNITED STATES

The concept of integration as a form of exchange between disciplines is not new. Originally called "liberal education," integration has been a goal of college education in the United States throughout the country's history (Chaves, 2014; Delbanco, 2012; Nussbaum, 1997; Zakaria, 2015). Study in an array of fields drawn from classical Greek and Latin roots constituted the sum total of college education from the Enlightenment to the era of industrialization. There were no major programs for undergraduates. Although today the term "liberal arts" is often used synonymously with "the humanities," this is not the original meaning of the term. A classic liberal education included training in grammar, logic, rhetoric (the Trivium), arithmetic, geometry, the theory of music, and astronomy (the Quadrivium), and today many schools include the natural and social sciences, along with the arts and humanities, in liberal arts degree programs. The founders believed that a broad and interwoven education was essential to the preparation of citizens for life, work, and civic participation. An educated and open mind would empower the college educated to separate truth from falsehood, superstition and bias from fact, and logic from illogic. The term "liberal" in the phrase "liberal education" has never meant "political liberalism." In Enlightenment usage and since, it has meant liberation of the mind[2] (Nussbaum, 2005).

Still, the proper relationship between academic fields is a long-standing issue in higher education. In the United States, higher education has had multiple goals since the second half of the nineteenth century, when federal legislation under the Morrill Act created the nation's land-grant universities to provide practical education in agriculture and the mechanical arts (Morrill Act, 1862). At the same time, higher education in the United States has functioned in a purposeful, distinctive, and cohesive way, uniquely emphasizing liberal education alongside practical education as nations elsewhere in the world have not done. Creation of land-grant universities did not supplant liberal education or its twentieth-century successor, general education. Indeed, the language of the Morrill Act states: "without excluding other scientific and classical studies and including military tactics, to teach such branches of learning as are related to agriculture and the

---

[2] See https://www.aacu.org/leap/what-is-a-liberal-education. (Accessed August 17, 2017).

mechanic arts, in such manner as the legislatures of the States may respectively prescribe, in order to promote the liberal and practical education of the industrial classes in the several pursuits and professions in life" (U.S.C. Title 7, Chapter 13, Subchapter I, §304). Liberal education, via general education, has run constantly through curricula of all types of institutions, from community colleges to the most elite universities.

In the late nineteenth century, as industrialization contributed to the rise of disciplines, the model of the German research university influenced the design of universities in the United States (Bonner, 1963; Diehl, 1978; Turner and Bernard, 1993; Wolken, n.d.). Where liberal education had previously constituted all of higher education, it now split from emerging specialized disciplines. Yet American higher education retained the goal of integration in the form of "general education," a term used to describe study in the arts and sciences—or synonymously, the "liberal arts." General education distribution requirements were intended to work by accretion, adding exposure to multiple art and science disciplines to the education of the major. Educators assumed that the integration would occur as students encountered a breadth of knowledge through general education together with the in-depth study of their majors (Gaff, 1991, pp. 32–63).

For over a century, then, curricula have generally sought to balance a specialized, practical education for workforce development with a general, liberal education that can contribute to an enlightened and engaged citizenry and a well-functioning democracy. As the influential Harvard report "General Education in a Free Society" put it, "the aim of education should be to prepare an individual to become an expert both in some particular vocation or art and in the general art of the free man and the citizen" (Conant, 1950). This goal has been embraced at all levels of higher education. Community colleges, for example, have come to recognize the value of a broad, liberal education for even the most practically oriented degrees (Albertine, 2012; David, 2015).

However, as both specialized majors and general education courses have come to be controlled by particular disciplines and departments, they have tended to evolve in the direction of specialization. Institutions of higher education both shaped and were shaped by this move toward increasing specialization. The expansion of federal investment in research since the end of World War II—and competition among academic fields for those resources—has contributed to the solidification of disciplinary boundaries and associated rationales for specialization in higher education.

In 1959 C. P. Snow famously lamented that the divisions between the sciences and the arts and humanities, which he described as "two cultures," were a rift of our own making. "The intellectual life of the whole of western society is increasingly being split into two polar groups," he wrote (Snow, 1959, p. 3). However, fields as disparate as aesthetics, ethics, mathematics, and natural philosophy have been conjoined domains of inquiry for at least two millennia. The rise of distinct disciplines is relatively recent. Not until the early nineteenth century was the term "scientist" coined by William Whewell to refer specifically to one whose vocation is inquiry into nature (Secord, 2014, pp. 105–107). It is something of an irony that Michael Faraday, who rejected this term and instead described himself as a "natural philosopher," is today almost universally referred to as a scientist (Secord, 2014, p. 105). The anachronistic application of the term is an index of how the idea that there are two separate and irreconcilable cultures is taken for granted.

When science was incorporated into liberal education in the late nineteenth and early twentieth centuries, it was done in the name of holistic education that would prepare young people for participation in civic life (Science, 2014). With the rise of increasingly segregated disciplines, this holistic sense of liberal education has attenuated: to use the metaphor given to us by Einstein, there has been greater emphasis on the individual branches than the tree as a whole. For instance, notions of "literacy" (e.g., of "science literacy" or "arts literacy") emphasize acquisition of basic competency within specific domains of knowledge, rather than affirming the older and, arguably, deeper and more demanding aspiration to educate individuals through exposure to the full sweep of human knowledge in the name of "liberating" them by engendering the capacities of critical reasoning, reflection, and engagement.

Scientific and technological specialization both benefited from and contributed to the increasing segregation of the disciplines. To give just one example, as engineering fields have expanded, spun off new subfields, and become progressively more specialized, undergraduate engineering degree requirements have grown in parallel (see "The Disciplinary Segregation of Higher Education" later in the chapter). With the emphasis on specialization and thorough coverage of an ever-burgeoning field of knowledge, undergraduate engineering majors enter the workforce with an expansive set of specific technical skills (Stephan, 2002). But, as a consequence, students have less space in their undergraduate careers to gain exposure to other fields and, thus, to other ways of seeing, understanding, and addressing problems, whether within or beyond engineering.

Similar trends are evident across many fields and majors, including the arts and humanities. Leon Botstein, president of Bard College and conductor of the American Symphony Orchestra, called for a "fundamental rethinking of professional training" back in 2000 (Botstein, 2000, p. 332). He acknowledged that more skilled and virtuosic musicians are alive now than ever in history, yet what is missing is not technique and expertise but interpretative and musical expression. He recommended that conservatories reorient their curricula toward a liberal education and interdisciplinary studies between music and other fields, such as mathematics, history, and psychology. Botstein posits that, by doing so, students will "deepen their musical skills and widen their curiosity and intellectual horizons."

As specialization has spread and majors have competed for students, programs of study in the majors have tended to add additional layers (and requirements), treating general knowledge (to say nothing of the integration of this knowledge) as the responsibility of the custodians of general education requirements. As general education lost ground to specialization over the past century, it has often taken the form of superficial exposure to a smattering of disciplinary approaches whose relationships and relevance to one another are rarely made clear to students (AAC&U, 2007; Association of American Colleges, 1994).

Even as this trend of increasing specialization was under way, shifts in U.S. demographics, dispute over the meaning and purpose of liberal education, and pressure from employers drove institutional leaders to propose curricular reforms, including reforms to general education. In 1991, Jerry G. Gaff wrote a careful analysis called "The Curriculum Under Fire" for *New Life for the College Curriculum* (Gaff, 1991) that laid out the grounds of the battle then being fought and argued for reform. In 1994, what was then the Association of American Colleges published *Strong Foundations: Twelve Principles for Effective General Education Programs* (Association of American Colleges, 1994), which made the case for the revival of general education because so much ground had been lost since the 1970s to the specialized majors. By the late 1990s, the Association of American Colleges and Universities (AAC&U) had launched a project titled Greater Expectations: A New Vision for Learning as a Nation Goes to College (AAC&U, 2002) and shortly thereafter published a national report by the same title that called for a "dramatic reorganization of undergraduate education to ensure that all college aspirants receive not just access to college but an education of lasting value (AAC&U, 2002, vii).

Today, public discourse continues to reflect a tension over whether higher education should primarily be considered a path to educated citizenship or employment, with surveys demonstrating that most Americans view higher education as a path to a "good job"(Gallup and Purdue University, 2014). This focus on preparation for employment has come to be associated with disciplinary specialization. As many people have come to see higher education more as a private commodity than a public good, states have invested less money in higher education and have increasingly linked resources to demonstrated employability of graduates (The Lincoln Project, 2015).[3] This has created pressure at many institutions to promote

[3] See https://www.amacad.org/content/publications/publication.aspx?d=21942. (Accessed August 17, 2017).

skills and training that are considered more likely to lead to immediate job placement, which has led to a perceived need for greater specialization. This trend has reinforced a view of higher education as largely a path for workforce preparation, contrary to the historical mission of institutions for higher learning to also prepare students for life and productive citizenship. At the same time, the value proposition of American higher education has become less apparent to a large percentage of the population (Gallup and Purdue University, 2015; Pew Research Center, 2011). Many Americans are struggling to understand the return on investment in a college education as they weigh the rising cost, which far outpaces inflation, with the fact that more and more employers are requiring a postsecondary degree for jobs that did not previously require one.[4]

But as other sections of this chapter will demonstrate, evidence suggests that employers, students, and proponents of liberal education now have greater agreement on many of the desired learning outcomes from higher education. Perhaps the tension between preparation for career, on the one hand, and preparation for life and civic engagement, on the other, is abating.

DISCIPLINARY INTEGRATION AND ISSUES OF EQUITY AND DIVERSITY IN HIGHER EDUCATION

> We want one class of persons to have a liberal education, and we want another class of persons, a very much larger class of necessity in every society, to forgo the privilege of a liberal education and fit themselves to perform specific difficult manual tasks.
>
> —Woodrow Wilson, 1909 Address to the NYC High School Teachers' Association

No discussion of the history of higher education would be complete without an acknowledgment of how historical inequities in access to higher education in the United States are reflected today. Women, people with disabilities, and population groups that include African Americans, Latinos, and indigenous peoples have not participated equitably in the arts, humanities, and science, technology, engineering, mathematics, and medicine (STEMM) fields.[5] As this committee considered the evidence of the benefits of integrative learning in the arts, humanities, and STEMM subjects in the aggregate, we likewise sought evidence of the benefits of integrative learn-

---

[4] See http://burning-glass.com/research/credentials-gap/. (Accessed August 18, 2017).

[5] Digest of Educational Statistics: Table 322.30. Bachelor's degrees conferred by postsecondary institutions, by race/ethnicity and field of study: 2013–14 and 2014–15. https://nces.ed.gov/programs/digest/d16/tables/dt16_322.30.asp. (Accessed August 18, 2017).

ing to groups of people who have been historically underserved by higher education. The committee contends that any new movement in higher education *must* ensure that it prepares *all* students to prosper economically, contribute civically, and flourish personally.

Improving the representation of women and minorities in STEM subjects is a national priority. In February 2012, the President's Council of Advisors on Science and Technology reported the following:

> Although women and members of minority groups now constitute approximately 70 percent of college students, they are underrepresented among students receiving undergraduate degrees in STEM subjects (approximately 45 percent). These students are an "underrepresented majority" that must be part of the route to excellence. Members of this group leave STEM majors at higher rates than others and offer an expanding pool of untapped talent. The underrepresented majority is a large underutilized source of potential STEM professionals and deserves special attention. (PCAST, 2012, p. 5)

Issues of equity and diversity in higher education intersect with the goals of disciplinary integration in two major ways. First, the committee found instances in which, either implicitly or explicitly, the goals of disciplinary integration are to make connections between STEMM fields and other disciplines so that STEMM subjects (and hopefully STEMM careers) become more appealing to groups traditionally underrepresented, and at times actively excluded, from STEMM fields. Because traditional STEMM pedagogies have been shown to discourage women and people of color from pursuing STEMM majors and careers (Byars-Winston et al., 2010; PCAST, 2012), some courses and programs that integrate the humanities, arts, and STEMM fields aim to expand the pedagogical repertoire of STEMM courses and programs. In Chapter 6 of this report, we review the limited, though encouraging, evidence on the impact of such efforts on women and underrepresented minorities in STEM.

Second, the committee observed that the topic of integration relates to arguments for the value of cognitive diversity. Cognitive diversity is characterized by the human capacities to form an array of perspectives and to take different cognitive approaches within fields of intellectual endeavor. Engaging diverse intellectual and cognitive perspectives has distinct benefits (Hong and Page, 2004). Differing minds and differing backgrounds can offer a wider variety of angles on a problem and can predict a wider array of outcomes than is likely to emerge from a homogeneous group or perspective. Economist Scott Page and colleagues have demonstrated that groups of cognitively diverse agents are better at problem solving and make more accurate predictions than individuals or homogeneous groups (Hong and Page, 2004). Extrapolating on the research of Page and others, some argue that an integrative approach to education has value because it can facilitate cognitive diversity among individuals and groups. A person with an integrative education is likely to have experiences working with a diversity of persons and a diversity of ideas and mindsets, which has been associated with positive outcomes. Diversity in higher education encourages critical thinking, challenges practitioners to consider multiple points of view; promotes teamwork across disciplinary boundaries; encourages an appreciation of differences in background, training, and aptitudes; and highlights the importance of the contexts in which education, work, and citizenship develop and are displayed (Allen et al., 2005; Bowman, 2010; Curşeu and Pluut, 2013).

In theory, disciplinary integration demands similar forms of boundary crossing. It invites students to learn with, across, and among the many people who teach and study at the intersections, and at the margins, of disciplines and fields. It promotes interplay and intersection among heuristics and epistemologies. Attention to diversity shapes the understanding of integration and interdisciplinary work. It increases the odds that the following questions will be asked: Who is defining the problems? Who is devising their solutions? How are the benefits and potential disparities of solutions distributed across groups?

However, in order to ensure equitable access to any learning experience, including integrative learning experiences, it is important to acknowledge the influence of the rising cost of college on who in our society may benefit from higher education and from which institutional types. As the Woodrow Wilson quote that opens this section of the report acknowledges, underserved groups have historically been driven by financial and societal pressures into more workforce-oriented educational pathways. Fortunately, as this report will demonstrate, the committee found that integrative models of higher education have emerged at almost all institutional types, including many of the community colleges and technical schools that serve a large proportion of the nation's underserved and underrepresented students (see Chapter 6). To the extent that integrative approaches to higher education are beneficial to students, the adoption of these approaches to education by a wide range of institutional types is an important step toward ensuring that these benefits are shared equitably.

THE DISCIPLINARY SEGREGATION OF HIGHER EDUCATION

Today there is a widely held belief that colleges and universities have become "siloed" along disciplinary lines such that faculty from different disciplines rarely interact, and students are thought to take most of their courses within the discipline of their declared major. Though existing data on the courses students are taking today in higher education are not granular enough to permit an analysis of the extent of this siloing, certain internal and external pressures are placed on universities that, the committee observed, are likely to drive disciplinary segregation and serve as barriers to integration.

Take, for example, the external influence of professional societies and the accreditation process. Professional organizations such as the American Chemical Society and the American Physical Society develop standards for certified degree programs, which in turn influence how accreditation bodies appraise degree programs in these disciplines. Accrediting bodies such as the Accreditation Board for Engineering and Technology (ABET) and the Higher Learning Commission have specific standards for accreditation of programs based on disciplinary constraints. For instance, ABET lists accreditation standards for 28 different categories of engineering, many of which have multiple subcategories defined as well. Institutions build their expectations for fully accredited programs based on these standards, which influences the makeup of the curriculum. In some cases, this has the effect of narrowing a student's curriculum along disciplinary lines. For example, engineering programs often include so many required courses that there is very limited space in the curriculum for students to study in disciplines outside of engineering (ABET, 2018).

Accreditation requirements also contribute to disciplinary segregation by influencing teaching practices. A recent change to faculty qualifications from the Higher Learning Commission requires faculty to have a minimum of 18 graduate-level hours in the specific content of a subdiscipline in order to be considered qualified to teach.[6] It is unclear how this standard would be applied to faculty teaching an integrated course.

It is also important to note how structural features of certain colleges and universities shape or restrict course-taking. At many large universities, students apply to a college within the university (such as the college of business, college of engineering, college of arts and sciences, college of education, etc.), which has its own curriculum and course distribution requirements. While it may be possible to take courses in the other colleges, the initial choice of a college within the university effectively determines the courses available to a student. Although there is some ability to change colleges within large universities, the initial choice of college restricts the kinds of courses that one can take to a much greater extent than, for instance, in a liberal arts college, where the selection of a major can come later and there is greater freedom to explore different fields.

Within institutions, traditional disciplinary control of faculty positions and of promotion and tenure requirements can also be a significant bar-

---

[6] See http://download.hlcommission.org/FacultyGuidelines_2016_OPB.pdf. (Accessed August 18, 2017).

rier to integration. Because tenure requirements tend to be determined and enforced by the disciplines, the expectations set for tenure and promotion by discipline-based faculty may be ill suited to evaluate faculty who work between and across disciplines. For example, the professional pressure to publish in peer-reviewed, high-profile journals often prevents faculty from publishing their work in journals of interdisciplinary studies, which are often considered less prestigious, even if their work might be a good fit for such publications (Jacobs and Frickel, 2009). Rather, faculty will feel pressure to prioritize research and scholarship that can be published in the high-impact journals associated with specific disciplines. Further, an individual's ability to be considered for jobs at other colleges or universities is largely determined by the ease with which a department can recognize the worth and value of prior research done by that person, this often made more difficult if an applicant's work straddles multiple departments (Jacobs and Frickel, 2009). In addition, teaching assignments and expectations often cover the disciplinary teaching first, and time and effort may not be permitted for more innovative, integrated teaching that falls outside the discipline/department. Even when integrated teaching and research are permitted, faculty rooted in more traditional disciplines may not ascribe as much value to those activities. While there may not be overt discrimination, the pressure to conform may make the integrative faculty member uncomfortable and less inclined to pursue scholarship and teaching across disciplines (Calhoun and Rhoten, 2010; Jasanoff, 2010).

In addition, decentralized budget models, such as responsibility-centered management, where funding for teaching flows primarily to disciplines and specific colleges, promote greater disciplinarity in higher education and make it difficult to develop and maintain interdisciplinary, transdisciplinary, and integrated courses and programs. Crossing boundaries of disciplines requires a conscious effort by administrators to provide incentives and support for integration (Wilson, 2002).

Other stakeholders also have input into the process of planning new programs and curricula. Besides the obvious faculty presence in program development, colleges and universities must answer to advisory boards and governing boards as they consider changes to programs. Movement toward more integrative programs will likely require that program planners communicate to governing bodies the relationship between an integrative program and the learning outcomes and skills students need to be successful in an evolving world. Similarly, the shift toward more integrated curricula will demand multistakeholder discussion of the criteria for accreditation.

At community colleges, there is an additional level of pressure to silo the disciplines that comes from the federal financial aid process, particularly the Pell Grant, and its effect on the community college curriculum.[7] There has been a strong effort to put policy in place to ensure that a greater number of students who begin their degree program, whether it be a 2-year associate of applied science (AAS) degree or a 2+2 associate of science (AS) transfer program, will complete the program within a reasonable time period. While some believe that these regulations were devised to discourage "Pell runners" (students who move from major to major or college to college to maximize the eligibility for financial aid that is not likely being used to support their education), there has also been a concern that poor advising—at the family, high school, and college levels—has led to too many community college students "wasting" credits by taking courses that do not transfer to a 4-year program. For these reasons, continued eligibility for financial aid requires that a student makes satisfactory academic progress, which entails specific timeline restrictions—currently that a student finish their program within 150 percent of the program plan (e.g., by 6 years for a 4-year bachelor's degree). In order to ensure that these students have the greatest potential to complete their program, many community colleges are requiring students to declare their program of study immediately upon matriculation, after which point students receiving aid can take only courses that are listed as part of the program. These restrictions, while intended to improve student success, have the unintended consequences of both inhibiting student exploration and limiting any creativity in the community college curriculum. Unless an integrated course has been carefully mapped to the learning goals of the disciplines integrated, and unless the receiving institutions have predetermined that an integrated course meets their requirements for general education, these kinds of non-discipline-specific courses would not be accessible to students receiving federal aid.

Additionally, more and more states are moving to a community college transfer model in which the community college is set up to mirror the first 2 years of the state 4-year baccalaureate programs. For example, the 2-year University of Wisconsin (UW) Colleges have been part of the University of Wisconsin System since 1971, granting associate of arts and sciences degrees and an array of transfer pathways to the UW universities. And in Connecticut, the students who take a Transfer Ticket degree program can transfer to one of the Connecticut State Universities without losing credits or having to take additional credits. The Texas public college and university system is another example. Texas requires the 42-hour Texas Core Curriculum (TCC) for students at all public institutions. TCC core objectives

[7] See https://www.fastweb.com/financial-aid/articles/can-a-student-be-cut-off-from-financial-aid-after-taking-too-many-credits. (Accessed August 18, 2017).

and component areas are shared and aligned across institutions, a design intended to facilitate transfer.[8]

In order for such programs to prepare students for success at the transfer institution, and to maintain levels of quality in learning, community college courses are often aligned or matched with courses at state universities, sometimes using the same course number and syllabus. These programmatic and financial aid structures have been created with the best of intentions, to create the best possible scenario for program completion. But by their very nature, they limit options for interdisciplinary course and curriculum design and unintentionally support a more siloed structure that is easy to match and monitor.[9]

However, while these strict financial aid and program structures limit the opportunities to integrate STEM and the arts and humanities in the 2 + 2 transfer programs, they actually have the potential to open up opportunities in the terminal two-year AAS programs. In an effort to streamline the general education component of these degrees, these types of integrative classes are often already a component of those programs. For example, in the AAS degrees at Mid Michigan Community College, students are required to take three 200-level integrative general education courses—Integrative Science (SCI 200), Integrative Humanities (HUM 200), and Integrative Social Sciences (SSC 200)—in addition to math and writing courses. While these courses are merely integrated within the general discipline, for instance the Integrative Science course content includes biology, chemistry, earth science, physics, and environmental science, along with a healthy dose of public policy and economics, they have the potential to become much more generally integrative. Because many of the AAS curricula have very little room for additional credits (e.g., the nursing program and the physical therapist assistant (PTA) programs are already longer than 2-year programs), the general education component of the curricula is often forced into a less discipline-specific mode so that students can have the breadth of the general education experience but in a smaller amount of time and within the framework of their program. Community college faculty are starting to look at these time and content challenges as opportunities to create new approaches to general education that more completely integrate the sciences, social sciences, and humanities.[10]

---

[8] See http://www.uwc.edu/; http://www.ct.edu/transfer/tickets; http://www.thecb.state.tx.us/index.cfm?objectid=417252EA-B240-62F7-9F6A1A125C83BE08. (Accessed August 18, 2017).

[9] See https://www.tccns.org/; https://www.cccs.edu/educator-resources/common-course-numbering-system/; https://www.accs.cc/index.cfm/workforce-development/career-technical-education/course-directory/. (Accessed August 18, 2017).

[10] See https://www.midmich.edu/program-guidesheets-2017-18/AAS%20Automotive%20Technology%20-%202017-2018.pdf. (Accessed August 19, 2017).

THE NEED FOR GRADUATES TO BE ADAPTABLE, LIFELONG LEARNERS

> The illiterate of the 21st century will not be those who cannot read and write, but those who cannot learn, unlearn, and relearn.
>
> —Alvin Toffler, American Writer and Futurist

Collaborative, critical thinking, and communication skills are valuable in an enormous range of professional domains, particularly in an era where jobs are rapidly changing. One could argue that today, more than ever, graduates need to be adaptable and lifelong learners. Memorization and long-term retention of knowledge hold less of a premium when all content knowledge is ostensibly accessible in the mobile devices in our pockets. In addition, students need to learn how to learn. They need to learn how to find information, analyze it for its validity, understand its application in different circumstances, and communicate it clearly and accurately to others. They need the critical thinking, logical reasoning, and lifelong learning attitudes required to determine whether a news headline on social media is fake and misleading or whether it offers valid and useful information upon which to base a decision. These skills and abilities will serve graduates not only in their lives as citizens and individuals but in their professional pursuits.

Interestingly, U.S. Census Bureau data show that STEM majors and arts, humanities, and social sciences majors often end up in professions that are not directly aligned with their major, and that specific occupations attract students with multiple kinds of academic preparation (see Figure 2-1). These data raise questions about how well a college or university curriculum focused on a specific, disciplinary major will serve students after graduation.

Moreover, graduates should be prepared not only to take a job that does not directly relate to their college major but also to change jobs and careers often throughout their working years, particularly in the years just after graduation. According to a 2016 report from the U.S. Department of Labor, the median number of years that younger workers (ages 25–34 years) stayed in a single job was 4.2 years (Bureau of Labor Statistics, 2016). These data suggest that graduates will be well served by skills and competencies that are transferrable from one job to another, as well as by the ability to be adaptable, lifelong learners who can pick up the new knowledge they may need for each new job.

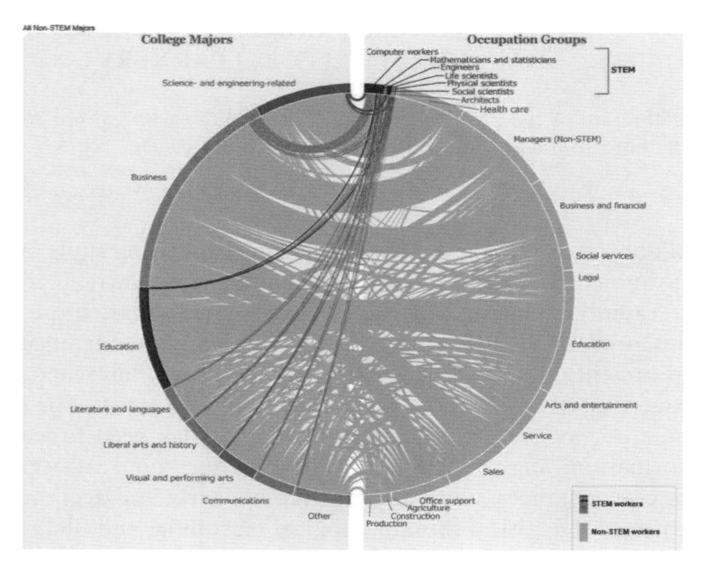

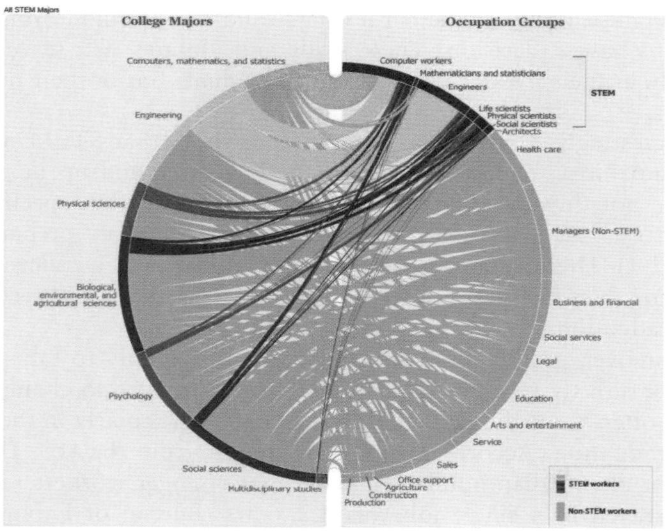

FIGURE 2-1 Students' college majors, whether in a STEM discipline or a non-STEM discipline (arts, humanities, and social sciences), do not necessarily predict their occupations.
NOTE: In this illustration, the length of each circle segment indicates the proportion of people graduating in each college major and employed in each occupation group. The thickness of the lines between majors and occupations indicates the share of people in that major-occupation combination.
SOURCE: https://www.census.gov/dataviz/visualizations/stem/stem-html/.

INTEGRATION AND HUMAN LEARNING

The need for lifelong learning skills and the ability to view issues from a variety of perspectives is also consistent with emerging evidence about how people learn and how expertise is developed (National Research Council 2000). Research shows that humans construct knowledge and understanding based on previous learning and experience. Evidence from the learning sciences clearly demonstrates that moving learners from being novices to experts requires more than an increase in content knowledge. While this storehouse of knowledge is certainly necessary for building expertise, it is far from sufficient. Beyond possessing a reservoir of content knowledge, experts also have the ability to categorize and sort information, to more readily relate and connect information that novices view as disparate, and to relate that information to newly encountered content, skills, or concepts in ways that novices are incapable of doing. Learning new information is often easier and more rapid for experts when that information fits logical patterns that they are able to construct cognitively. Experts thus build what learning scientists refer to as conceptual frameworks, which allow them to think more deeply about relationships among pieces of information that have already been learned and to better envision how seemingly disconnected information fits with, and is related to, what is already understood. Indeed, helping nonexperts learn by encouraging and actively teaching them how to develop conceptual frameworks may also enable them to learn content more readily because they can then better understand the relevance of that information and its connections with otherwise seemingly disparate facts.

Too often in undergraduate education, expectations for learning, and assessments of learning focus on acquiring a body of content that students see as disconnected or irrelevant to their interests. In this conventional approach to undergraduate education, the "big picture" that ties all of these facts and concepts together may be introduced toward the end of the course (if time permits) or sometimes toward the end of the student's tenure in college as some kind of "culminating experience." But proponents of integration argue that helping students integrate and understand the big picture connections between the different courses they take should be a more consistent feature of higher education because it is likely to make learning easier, more relevant, and more engaging. They posit that the multiple ways in which the arts, humanities, and STEMM fields have developed to understand and answer problems can be harnessed to expand the breadth and power of conceptual frameworks. The natural sciences have already recognized the power of seeking answers to large questions through the integration of multiple disciplines[11] (Labov et al., 2010; National Research Council, 2009)—albeit within the STEMM disciplines—and, based on the evidence cited in this report, it appears that these principles of learning could potentially be used to expand *all* students' understanding of complex problems through more deliberate integration of the arts, humanities, and STEMM.

Research on how students learn has also demonstrated that well-designed project- and inquiry-based approaches, often called "experiential learning," increase student learning (Brown et al., 1989; Kuh, 2008). Project-based learning situated in real communities, where students have opportunities to work on open-ended, integrative problems, has been shown to have long-term positive impacts on students, including confidence in their STEM courses, confidence in their jobs, and confidence in their capacity to solve complex problems (Vaz and Quinn, 2014). Significantly, research has found that novelty, unpredictability, and mutability in projects enhances learning, growth, and confidence, and that women and students of color more readily embrace STEM subjects and questions when the material is contextualized (Brunner, 1997; Margolis et al., 2000; Tobias, 1993; Tsui, 2007). Though not all courses and programs that embrace experiential learning integrate the humanities, arts, and STEMM subjects, many do. This may be due, in part, to the fact that many real-world problems have dimensions that are humanistic, scientific, technical, medical, and aesthetic.

Responding to the new scientific knowledge on the ways that people learn, many leaders of general education on campuses are seeking to achieve student learning outcomes that are integrative (see the section of Chapter 6 on within-curriculum integration). They are shifting emphasis from knowledge being provided passively to students through "teaching by telling" to the capacities or abilities that students gain and demonstrate as they learn (Barr and Tagg, 1995). At the same time, integrative and interdisciplinary designs have become an area of focus for college education (Klein, 2010).

THERE IS BROAD AGREEMENT BETWEEN EMPLOYERS AND INSTITUTIONS OF HIGHER EDUCATION ON CERTAIN STUDENT LEARNING GOALS

Surveys of employers and institutions of higher education have demonstrated that there is broad agreement on the types of learning outcomes that all graduates should leave higher education having achieved (Burning Glass Technologies, 2015; Hart Research Associates, 2016). Among those outcomes called for by both employers and higher education institutions are writing and oral communication skills, critical thinking and analytical

[11] See https://www.nap.edu/catalog/18722/convergence-facilitating-transdisciplinary-integration-of-life-sciences-physical-sciences-engineering. (Accessed August 19, 2017).

reasoning skills, teamwork skills, ethical decision making, and the ability to apply knowledge in real-world settings.

Shared Goals for Student Learning Among Diverse Institutions of Higher Education

Though higher education in the United States is incredibly diverse, comprising a vast array of different types of institutions that serve a variety of student goals and educational purposes, research has demonstrated that there is broad agreement across this diverse landscape of institutions on certain student learning outcomes. A survey released in 2016 found that nearly all AAC&U member institutions—which constitute a majority of 4-year colleges and universities in the United States—have adopted a common set of learning outcomes for all their undergraduate students (Hart Research Associates, 2016). Shared learning outcomes included writing and oral communication skills; critical thinking and analytical reasoning skills; knowledge of science, the humanities, the arts, mathematics, the social sciences, and global world cultures; ethical reasoning skills; and "integration of learning across disciplines," among others (Figure 2-2). The survey also found that general education is growing as a priority and that administrators are more likely than they were in 2008 to report an emphasis on the integration of knowledge, skills, and applications. This integrative—or reintegrative—turn of higher education signals a shift to intentional and purposeful learning across knowledge, skills, and personal and social responsibility (National Leadership Council for Liberal Education and America's Promise, 2007).

In 2005, the AAC&U launched its Liberal Education and America's Promise (LEAP) initiative (Figure 2-3). The Essential Learning Outcomes of the LEAP initiative are well regarded and frequently used to design undergraduate education. These goals give priority to application, integration, and high-impact learning and emphasize student learning outcomes in inquiry and analysis, critical thinking, teamwork, written and oral communication skills, and ethical reasoning and action, among other learning outcomes (see Figure 2-3). Preparing students for a world of unscripted problems is the goal (AAC&U, 2015). More recently, the LEAP challenge, launched during the AAC&U's centennial in 2015, invited institutions to make "signature work" a goal for all students.[12] Signature work asks students to pursue a significant project of their choosing as part of their college education. Nearly all employers surveyed by AAC&U (89 percent) said that they want to hire graduates who have skills and experience of this kind.[13]

---

[12] See https://www.aacu.org/leap-challenge. (Accessed August 19, 2017).

[13] See https://www.aacu.org/leap/public-opinion-research/2015-survey-falling-short. (Accessed August 19, 2017).

Proportion of Institutions That Have Learning Outcomes for All Students That Address Specific Skills and Knowledge Areas *(among institutions that have a common set of learning outcomes for all students)*[1]	2008 %	2015 %
Writing skills	99	99
Critical thinking and analytic reasoning skills	95	98
Quantitative reasoning skills	91	94
Knowledge of science	91	92
Knowledge of mathematics	87	92
Knowledge of humanities	92	92
Knowledge of global world cultures	87	89
Knowledge of social sciences	90	89
Knowledge of the arts	N/A	85
Oral communication skills	88	82
Intercultural skills and abilities	79	79
Information literacy skills	76	76
Research skills and projects	65	75
Ethical reasoning skills	75	75
Knowledge of diversity in the United States	73	73
Integration of learning across disciplines	63	68
Application of learning beyond the classroom	66	65
Civic engagement and competence	68	63
Knowledge of technology	61	49
Knowledge of languages other than English	42	48
Knowledge of American history	49	47
Knowledge of sustainability	24	27

FIGURE 2-2 Surveys of Association of American Colleges and Universitities member institutions demonstrate a growing commitment to common learning outcomes across institutions.

Certain Student Learning Outcomes Are Broadly Valued by Employers

Given the need for innovation in modern economies, employers know that a variety of employee talents are essential to the competitiveness and growth of their organizations. But recent surveys of employers reveal that they see talent as more than deep technical expertise or familiarity with a particular technology. They are also looking for well-rounded individuals with a holistic education who can comprehend and solve complex problems embedded within sophisticated systems that transcend disciplines;

The Essential Learning Outcomes

★ ★ ★ ★ ★ ★ ★ ★ ★ ★ ★ ★ ★ ★ ★ ★ ★ ★

Beginning in school, and continuing at successively higher levels across their college studies, students should prepare for twenty-first-century challenges by gaining:

★ Knowledge of Human Cultures and the Physical and Natural World

- Through study in the sciences and mathematics, social sciences, humanities, histories, languages, and the arts

Focused *by engagement with big questions, both contemporary and enduring*

★ Intellectual and Practical Skills, including

- Inquiry and analysis
- Critical and creative thinking
- Written and oral communication
- Quantitative literacy
- Information literacy
- Teamwork and problem solving

Practiced extensively*, across the curriculum, in the context of progressively more challenging problems, projects, and standards for performance*

★ Personal and Social Responsibility, including

- Civic knowledge and engagement—local and global
- Intercultural knowledge and competence
- Ethical reasoning and action
- Foundations and skills for lifelong learning

Anchored *through active involvement with diverse communities and real-world challenges*

★ Integrative and Applied Learning, including

- Synthesis and advanced accomplishment across general and specialized studies

Demonstrated *through the application of knowledge, skills, and responsibilities to new settings and complex problems*

FIGURE 2-3 The essential learning outcomes of the Association of American Colleges and Universities' Liberal Education and America's Promise initiative emphasize the competencies that students need for work and civic participation in the 21st century.
SOURCE: Association of American Colleges and Universities.

understand the needs, desires, and motivations of others; and communicate clearly.

An online survey conducted on behalf of the AAC&U found that the majority of employers[14] say that both field-specific knowledge and a broad range of other kinds of knowledge and skills are important for recent college graduates to achieve long-term career success (Hart Research Associates, 2013, 2016). Very few employers indicate that acquiring the knowledge and skills needed primarily for a specific field or position is the best path to long-term success. Employers report that, when hiring, they place the greatest value on demonstrated proficiency in skills and knowledge that cut across all majors. The skills that they rate as most important include the ability to communicate clearly, both in writing and orally, teamwork, ethical decision making, critical thinking, and the ability to apply knowledge in complex, multidimensional, and multidisciplinary settings. According to employers, this combination of cross-cutting skills is more important to an individual's success at a company than the major he or she pursued while in college. Similarly, a survey of Massachusetts Institute of Technology (MIT) alumni demonstrated that graduates rely more heavily on communication, teamwork, and interpersonal skills throughout their careers than the specific technical and engineering skills that they learned as undergraduates (Box 2-1).

A study conducted by Burning Glass, a job market analysis company, reported similar results. Its textual analysis of 25 million job postings aimed at understanding "the essential or baseline skills that employers are demanding across a wide range of jobs" revealed that oral communication, writing, customer service, organizational skills, and problem solving were among the most high-demand skills across a wide range of occupation and career types (Burning Glass Technologies, 2015). The study also categorized the importance of what were termed "baseline skills" and "technical skills" by occupation groups. The results of this analysis speak to the importance of *both* job-specific technical and baseline skills and the relative importance of these skills by occupation type. Jobs categorized as being within the domains of information technology, engineering, health care, the physical and life sciences, mathematics, and manufacturing require more technical skills than jobs such as sales, marketing, or human resources. However, even among the highly technical fields a quarter to a third of the required skills deemed essential by employers fall within the baseline skills. The results suggest that

[14] The 318 employers surveyed by Hart Research Associates were executives at private-sector and nonprofit organizations, including owners, CEOs, presidents, C suite–level executives, and vice presidents, whose organizations have at least 25 employees and report that 25 percent or more of their new hires hold either an associate's degree from a 2-year college or a bachelor's degree from a 4-year college.

BOX 2-1
MIT Alumni Survey Shows Graduates Use Communication Skills More Than Technical Skills in Their Careers

A study conducted by K. Wang (MIT, 2015) found that much of the technical knowledge mechanical engineers learned during their undergraduate preparation was not used in practice when they entered the workforce. To carry out the study, Wang collected data through a survey of the graduating classes of 1992 through 1996, 2003 through 2007, and 2009 through 2013. The survey focused on the frequency of use, expected proficiency, and source of knowledge pertaining to technical knowledge, engineering skills, work environment skills, and professional attributes. The responses to the survey "show a lower frequency of use for the technical reasoning knowledge and a high frequency of use for communication-based skills." The author concludes that this is "because technical knowledge is considered valuable to a specialized group of people, whereas the work environment skills are more career-independent." These data suggest the need for a rebalancing of subject matter and pedagogical approaches.

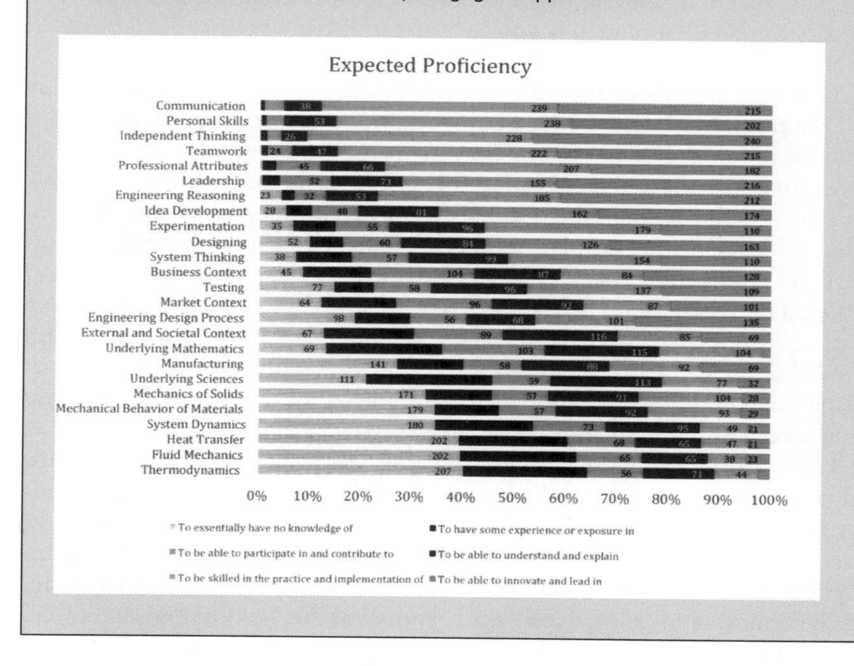

higher education should equip *all* students with the baseline skills needed for success in a wide range of occupations and, to the extent possible, cater specific technical instruction to the student's intended career path (but, as noted in Figure 2-1, students are likely to actually be employed in many sectors besides what they studied in the college major).

Employers also reported that many recent college graduates have not achieved the kinds of learning outcomes that they view as important. This is especially the case for applying knowledge and skills in real-world settings, critical thinking, and written and oral communication. In these areas, less than 30 percent of employers think that students are well prepared. Greater than 80 percent of employers feel that colleges and universities need to improve in helping graduates gain cross-cutting skills and knowledge. Large majorities of employers also indicate that various types of applied and engaged learning experiences—such as a comprehensive senior project, a collaborative research project, a research-based or applied field-based experience with people from other backgrounds, or a community-based or service learning project—would help applicants build the cross-cutting skills they are seeking and thereby positively influence their hiring decisions. A Gallup poll of U.S. business leaders from 2014 confirms these findings, concluding that both the applied skills and the amount of knowledge a job candidate has in a field are more important factors in hiring decisions than where a candidate attended school or the candidate's major (Calderon and Sidhu, 2014).

IBM has described the image for the kinds of workers employers say they want as the "T-shaped" employee—someone who combines depth of understanding in a particular field with the broad set of competencies needed to apply that understanding (IBM, 2009). Though this is a compelling and popular image that is consistent with an integrative approach to higher education, the problem with this image, in the committee's opinion, is its implication that depth and breadth are somehow separable and orthogonal, whereas in fact they work together seamlessly in the most productive and innovative employees. The committee envisions integration as more akin to a caduceus than a "T". In the caduceus, the breadth of learning that comes from exposure to multiple forms of knowledge continually interacts with and supports the development of depth.

Other groups of employers conceive of the need for a range of skills and competencies as "capability platforms." A task force of the organization STEM Connector, consisting of more than 30 leaders from industry, government, education, and the nonprofit sectors, identified four capability platforms that members of the STEM workforce need to be successful: digital fluency, innovation excellence, "employability" skills, and discipline-specific skills (STEM Innovation Task Force, 2014). The task force defined "employability skills" as a discipline-independent set of competencies and behaviors that all employers expect from their employees (STEM Innovation Task Force, 2015). These skills include teamwork, communication, reliability, and flexibility (the ability to understand and adapt to new ideas). Career-focused experiential learning—including problem-based learning, internships, team competitions, and all forms of work experience—is an especially effective way to build these skills, the task force observed, and this problem-based learning is often interdisciplinary in nature. As one business leader quoted in the task force's report said, "When you look at the types of problems that are out there today, these require people from different disciplines to come together to be able to solve them" (p. 17). This is not to say that technical skills are unimportant, rather that technical skills alone are insufficient.

Developing the broad set of skills desired by employers requires more than intensive study in a particular discipline during college. It requires exposure to multiple fields, practice to build employability skills, and experience with communication and collaboration. When the American Historical Association convened focus groups of Ph.D. historians who had found jobs beyond traditional academic positions, they agreed that the following five skills are necessary to be successful:

1. *communication*, in a variety of media and to a variety of audiences
2. *collaboration*, especially with people who might not share your worldview
3. *quantitative literacy*, a basic ability to understand and communicate information presented in quantitative form
4. *intellectual self-confidence*, the ability to work beyond subject matter expertise, to be nimble and imaginative in projects and plans
5. *digital literacy*, a basic familiarity with digital tools and platforms

No one academic discipline is most adept at cultivating all five of these skills.

Likewise, findings from the Strategic National Arts Alumni Project (SNAAP) survey, which is administered to more than 140,000 arts alumni nationally, indicates that students are receiving unparalleled training in art techniques but arts training is also encouraging experimentation, creativity, critical thinking, and problem solving. Recent alumni who responded to the SNAAP survey articulated many ways that this approach to arts training assists them in their work lives and contributes to their health and well-being, their relationships with others, their ability to collaborate and provide constructive criticism, and their ability to creatively solve problems. Arts graduates often see themselves as leaders at work and in their communities. Further,

> although arts graduates are warned that they will struggle to find employment after graduation and that their employment may not make use of their skill set, many graduates find work in the discipline of their training. On average, almost 7 of every 10 currently employed arts graduates described their current jobs as "relevant" or "very relevant" to their training (specifically, 64 percent of recent alumni and 69 percent of all alumni)—a

greater percentage than graduates from journalism, accounting, or biology majors. (Strategic National Arts Alumni Project, 2014, p. 23)

As noted earlier, many of these skills overlap with the "essential learning outcomes" identified by AAC&U's LEAP initiative, including critical thinking, teamwork and problem solving, quantitative literacy, written and oral communication, and integrative and applied learning. These are outcomes that employers and educators, including faculty from the arts, humanities, and STEM fields, agree that all students should gain from higher education (see Figure 2-3) (AAC&U, 2007).

Hiring patterns also lend support to the idea that employers value a broadly based education. A study by LinkedIn revealed that the growth of employees with undergraduate degrees outside of science and engineering who entered the technology sector between 2010 and 2013 exceeded computer science and engineering majors by 10 percent (Ma, 2015). Today, about 10 percent of graduates with majors other than science and engineering are going into the technology sector, including about one in seven of those who graduate from the top 20 colleges and universities in the United States (Ma, 2015).[15]

In summary, the skills, knowledge, and abilities that employers want are consistent with the goals of many institutions of higher education. Chapter 6 presents evidence that certain integrative approaches are associated with student outcomes that are consistent with these shared learning goals, including higher-order thinking, content mastery of complex concepts, enhanced communication and teamwork skills, and increased motivation and enjoyment of learning.

WHAT DO STUDENTS WANT?

On some measures, polls of what students want from higher education correspond with employers' desires. According to an online survey of 613 college students—all of whom were ages 18 to 29 and within a year of obtaining a degree, or in the case of 2-year college students, within a year of obtaining a degree or transferring to a 4-year college—more than four in five students say that doing well in their college studies and getting a good job are very important to them personally (giving the goals a rating of 8, 9, or 10 on a scale from 0 to 10). Students also know that they need knowledge and skills beyond those of a specific field or major to achieve success, with 63 percent indicating that both field-specific knowledge and a

[15] See https://blog.linkedin.com/2015/08/25/you-dont-need-to-know-how-to-code-to-make-it-in-silicon-valley. (Accessed August 21, 2017).

broad range of skills are important.[16] The skills and knowledge they identify as important include those that cut across majors, including the ability to apply knowledge in real-world settings, written and oral communication, teamwork, and ethical decision making (Hart Research Associates, 2016). Further, 84 percent of students say that thinking creatively is an important or very important skill to learn in college, and 92 percent say that a career that allows them to be creative is important (The Teagle Foundation, 2017).

On other measures, however, college students are notably out of sync with employers. Surveys demonstrate that university administrators and students feel confident that higher education is preparing graduates for the workplace, while most employers do not. According to a poll conducted by Gallup for the Lumina Foundation, 96 percent of chief academic officers at higher education institutions say their institution is "very or somewhat" effective at preparing students for the world of work (Busteed, 2014). However, only 14 percent of Americans strongly agree that college graduates are well prepared for success in the workplace, and barely 1 in 10 business leaders strongly agrees that college graduates have the skills and competencies that their workplaces need.

The interest of college students in a broadly based education is reflected in the courses they are taking. The Department of Education's National Center for Education Statistics reports a 37 percent increase in the number of students majoring in multi-/interdisciplinary studies between 2008–2009 and 2013–2014 (U.S. Department of Education, 2016). Another indication of growing interest in integrative education is the growing number of health humanities programs. From 2000 to 2016, the number of health humanities programs more than quadrupled, increasing from 14 to 57, with another 5 known programs currently in development. Increased enrollments in undergraduate public health programs, many of which pursue liberal arts outcomes, reveal a similar interest (Leider et al., 2015). Further, among the most popular majors at Stanford University is the major in science, technology, and society, through which students explore "what science and technology make of the world and what the world makes of science and technology" and take both technical courses in science and technology and courses in which they "study the social and historical context of science and technology, involving their global, ethical, political, organizational, economic, and legal dimensions."[17]

---

[16] See https://www.aacu.org/leap/public-opinion-research/2015-students (Accessed August 21, 2017).

[17] See https://sts.stanford.edu/about/what-sts. (Accessed August 21, 2017).

INTEGRATION AND THE INNOVATION ECONOMY

An approach to higher education that favors increasing specialization may not be well suited to today's challenges. When well-engineered technologies fail, for example, the root cause is often failures of empathy and imagination, not flaws in technical design (see Box 2-2). Conversely, the most successful products tend to marry mastery of technical design with functional or aesthetic insights about what people find useful, desirable, and beautiful.

In a world being transformed by technology, innovation is a key to economic success. As a recent report from the National Endowment for the Arts, 2017, p. 12) pointed out,

> The growth of global manufacturing competition has pressured companies to distinguish themselves on features other than price, volume, speed, and quality. In some industries, particularly in steel, cotton, tobacco, coal, and electronics, companies outside of the U.S. are able to outcompete American firms on these criteria, which motivates U.S. firms to find other ways of increasing their market share and corporate value. The ability to innovate has become a major differentiator, with industrial design as one of the key means by which companies are innovating today.

BOX 2-2
Cook Stoves and the Need for Integration

The world's poor have experienced a long-standing need for better cook stoves. Smoke from cooking and heating kills 4 million people a year and contributes to pollution and deforestation. Yet despite years of effort and a plethora of creative designs, engineers have not yet produced a technology that meets existing needs.

The failure, according to former United Nations deputy high commissioner for refugees T. Alexander Aleinikoff, lies in the disconnect between designers' imaginations of what makes for a good technology and the actual conditions—physical, social, and cultural—in which they will be used (Parthasarathy, 2016). "We're in the situation where everybody and his brother has invented a cookstove and none of them have really worked well for us" (Confino and Paddison, 2014). This failure traces, at least in part, to the segregation of the disciplines. The designer may see only the material and technical parameters of a technology and not the dimensions of lived experience and social reality that are the focus of the cultural anthropologist. This is an unsurprising outcome of an educational system that discourages rather than encourages engineers to see stoves as culturally embedded artifacts—that is, to attend to the human dimensions and context of an apparently technical problem.

Innovation today requires integrative thinking and collaboration, while technology development decreases the need to perform repetitive tasks and leaves more time for innovation and interdisciplinary collaboration among colleagues and occupations (Carnevale et al., 2011) (Box 2-3).

BOX 2-3
The Brighton Fuse Project

As technology continues to revolutionize modern society, innovations increasingly require a fusion of STEMM disciplines and the arts and humanities to create desirable, marketable, and accessible products. In the United Kingdom, Brighton, which is akin to Silicon Valley in California but with generally smaller, business-to-business firms, has become a hotspot for this type of creative innovation.

An early wave of innovators arrived in Brighton in the early 1990s and 2000s, drawn by the city's long history as an artistic center, its universities, and its attractive lifestyle. Eventually these founders created a critical mass that attracted others with more straightforward economic motivations. Most of the entrepreneurs in Brighton are well educated and older than their Silicon Valley counterparts, which might also contribute to their business success. Brighton's mixture of hard-nosed economics with softer cultural and institutional support has created a growing yet supportive, stable yet flexible network of businesses. By combining diverse skills and knowledge, these firms have an advantage over single-minded companies that are unable to adapt in a rapidly changing economy.

The Brighton Fuse Project has been studying the growth of successful creative and digital businesses in Brighton. Two-thirds of the firms are fused or superfused, meaning that they combine creative art and design skills with technological expertise. The project collected and analyzed qualitative and quantitative data from 1,495 local firms in the creative and digital industries to determine the nature of these interdisciplinary relationships. The Brighton Fuse Project discovered that the average fused/superfused firm was more innovative and grew at more than twice the speed of unfused firms, reflecting the changing nature of the global market and workforce. One of the key findings of the report was the prominence of arts and humanities skills, with 48 percent of Brighton entrepreneurs being arts, design, and humanities graduates. As knowledge becomes more specialized, it becomes harder to classify innovations into single categories and increasingly more important to coordinate skill sets to create the best product. Digital innovations rely on contributions from all disciplines, just as they are consumed by all different types of people.

The project has also revealed the importance of value capture in addition to value creation. Innovation might ignite growth, but this alone is not enough. The mechanisms through which innovations are released and distributed drive profit and determine success. As Brighton's reputation has grown, local firms have used its growing brand to their benefit. The diversity of firms in the environment also allows for more complex production processes than would be possible if the firms were working independently.

continued

BOX 2-3 Continued

Despite Brighton's success, the fusion effect does not occur automatically. Even within fused and superfused firms, respondents face major problems involving economic and financial barriers, a lack of visibility and bargaining power, and barriers related to skills, managerial gaps, and high workloads. The report concludes with three lessons from Brighton's success: (1) It is difficult to create clusters without a base on which to build, but policy can be helpful later. (2) Cluster development is a dynamic mix of hard-nosed economics and softer cultural and institutional support. (3) Creating interdisciplinary integration is difficult but can be done (Sapsed and Nightengale, 2013).

THE CORRELATION BETWEEN PARTICIPATION IN THE ARTS AND THE ADVANCEMENT OF SCIENCE, ENGINEERING, AND MEDICINE

Another rationale for the integration of disciplines is the notion that the integration of knowledge promotes innovative thinking that can lead to significant scientific breakthroughs. Evidence for this assertion comes in the form of strong correlations between participation in the arts and individual excellence in science, engineering, and medicine, as well as historical examples that document how breakthroughs in science have been inspired by analogies provided by the arts.

Like Einstein, many of the great minds in science, engineering, and medicine have subscribed to the idea that all knowledge is connected and have actively participated in the arts and humanities alongside their scientific pursuits. For example, the work of Robert and Michelle Root-Bernstein has shown very strong correlations between leadership in science and engagement with arts and crafts avocations. In a 2008 study, Bernstein and colleagues found that very accomplished scientists, including Nobel Laureates, National Academy of Sciences members, and Royal Society members, were significantly more likely to engage in arts and crafts and identify as artists than average scientists and the general public (Root-Bernstein et al., 2008). Compared with scientists who are members of Sigma Xi, a society in which any working scientist can be a member, Nobel Laureates were 2 times as likely to be photographers, 4 times as likely to be musicians, 17 times as likely to be artists, 15 times as likely to be crafts people, 25 times as likely to be creative writers, and 22 times as likely to be performers.

The Root-Bernsteins and colleagues have also found that sustained arts and crafts participation correlates with being an entrepreneurial innovator. A study that examined Michigan State University Honors College science and technology graduates from the period 1990–1995 found that (1) STEM majors are far more likely to have extensive arts and crafts skills than the average American, (2) arts and crafts experiences are significantly correlated with producing patentable inventions and founding new companies, (3) the majority believe that their innovative ability has been stimulated by their arts and crafts knowledge, and (4) lifelong participation and exposure in the arts and crafts yields significant impacts for innovators and entrepreneurs (LaMore et al., 2013). Further, a summit held in 2009 by the Neuro-Education Initiative of the Johns Hopkins School of Education reported that researchers found "tight correlations between arts training and improvements in cognition, attention, and learning."[18]

History is full of examples of people who drew upon their talent and passion for science and art to drive new discoveries and advances (Root-Bernstein and Root-Bernstein, 1999). In his book *Music and the Making of Modern Science*, Peter Pesic describes how breakthroughs in physical science and mathematics were inspired through musical analogies (Pesic, 2014). For example, Kepler's Third Law emerged from his search to describe the polyphony of the planets. Faraday discovered electromagnetic induction while investigating Wheatstone's novel musical/sonic devices, Newton imposed the musical scale on colors, and Helmholtz developed alternative geometric "spaces" in response to his work on music and vision. Interestingly, history also points to examples of artists contributing to scientific and technological breakthroughs. To cite just a few examples, composers Leopold Mannes and Leopold Godowsky invented the Kodachrome Color Film process, sculptor Patricia Billings invented "geobond" while trying to improve plaster, and artists Heather Ackroyd and Dan Harvey revolutionized plant nutrient screening through painting.

While these case studies and correlational studies are very interesting and do suggest a relationship between participation in the arts and scientific and entrepreneurial accomplishment, Bernstein and others point out that correlation should not be confused with causation. It might be the case that arts training makes for better scientists and entrepreneurs, but it might also be the case that Nobel Laureates who are concert pianists are simply extraordinary human beings, or alternatively, had more privileged and resourced upbringings than others.

We cannot say with any certainty that an education that integrates the arts with the STEMM subjects will necessarily lead to new foundational breakthroughs in science, engineering, and medicine. However, one can reasonably argue that an educational approach that teaches students to see the disciplines as distinct and nonoverlapping domains may discourage the

[18] See http://www.steam-notstem.com/wp-content/uploads/2010/11/Neuroeducation.pdf. (Accessed August 21, 2017).

boundary crossing that has been characteristic of some of humanity's most significant contributors to the advancement of knowledge. As such, curricula that cleave too tightly to a single discipline may hamstring the ability of students to think beyond the limits of what has already been thought, to achieve new forms of creative innovation, and thereby to better understand and address the challenges of the moment.

SHIFTING GROUND IN HIGHER EDUCATION

This chapter presented an overview of some of today's most pressing challenges and opportunities in higher education, including the following:

- *The need to achieve more effective forms of capacity building for twenty-first-century workers and citizens.* The nature of work requires that graduates acquire a broad base of skills from across the disciplines that can be flexibly deployed in different work environments across a lifetime. These skills overlap with the demands of citizenship, which require capacities to reflect upon and engage with questions of public import in an informed way. In a world where science and technology are major drivers of social change, historical, ethical, aesthetic, and cultural competencies are more crucial than ever. At the same time, the complex and often technical nature of contemporary issues in democratic governance demands that well-educated citizens have an appreciation of the nature of technical knowledge and of its historical, cultural, and political roles in American democracy.
- *The need to draw on the untapped potential for innovation and collaboration within and beyond the university.* The complex social, technological, and environmental problems of our historical moment demand creative solutions that are humane, technically robust, and elegant. Giving students broader repertoires for critical thinking and creative innovation has the potential not only to create a workforce and polity that can effectively confront these problems but also to uncover points of connection and synergy within and beyond academia that may prove generative. In short, a commitment to a more integrated education has the potential to push the "multiversity" itself toward reintegrating into a university.
- *The need to cultivate more robust cultural and ethical commitments to empathy, inclusion, and respect for the rich diversity of human identity and experience.* Truly robust knowledge depends on the capacity to recognize the critical limitations of particular ways of knowing, to achieve the social relations appropriate to an inclusive and democratic society, and to cultivate due humil-

ity. These commitments are as essential to productive professional environments as they are to wider civic life. They are also critical to creating shared aspirations of the futures we want and to cultivating the forms of inquiry, innovation, and creative expression that will help build those futures.

3

What Is Integration?

Higher education has not yet agreed on a definition of what the integration of the arts, humanities, and science, technology, engineering, mathematics, and medicine (STEMM) fields is and what it is not. As a result, many questions continue to surround integration. Is the use of poetry or song assignments in a science course integrative if literature or music theory faculty are not involved? What makes a course with equal parts sculpture and engineering more or less integrative than a course focused on engineering design? Music theory courses may cover the mathematics found in music, but does that make them integrative? When an educator claims an educational experience is integrative, does that mean it actually is?

Where in the higher education curricula should integrative approaches be adopted or experimented with? In general education? In the major? In co-curricular activities?

The committee found that there are many diverse approaches to integration. Different disciplines are integrated at different levels of depth and for different reasons. Different courses and programs use different pedagogical approaches and appear in different aspects of the curriculum. This chapter explores some of this diversity and concludes that there is no single goal of an integrative approach, but rather many different goals. The many and varied goals of integration have implications for how the impact of integration on students should be evaluated by institutions.

Below we offer definitions of the disciplines (arts, humanities, science, engineering, and medicine) that have been developed by others over the course of time and describe the characteristics of integration. We also describe forms of integration in the curriculum, broken into three categories:

in-course, within-curriculum, and co-curricular. We describe the differences between "interdisciplinary" integration, "multidisciplinary" integration, and "transdisciplinary" integration and acknowledge that "integration" is a term used in higher education research that may or may not refer specifically to the integration of the humanities, arts, and STEMM fields. Rather, "integration" in the context of the higher education scholarship may refer broadly to educational experiences that help students integrate or bring together ideas.

THE DISCIPLINARY CONTEXT

Integration of teaching and learning in higher education inevitably takes place within the context of disciplinary pedagogies, content, and epistemologies. Disciplines have their own ways of looking at the world, of making meaning and discovering truth. But these approaches are pragmatic, meant to arrive at certain human ends. The disciplines delimit their objects of study; their theoretical approaches, projects, and traditions; the forms of evidence, interpretation, and explanation that are appropriate to them; and the professional and institutional structures through which these parameters are articulated, regulated, taught, and, in effect, enforced. The disciplines are self-reinforcing, and disciplinary specialization and fragmentation have intensified as the disciplines have strengthened and solidified.

As historians of higher education and of integrative learning have long observed, the disciplines have their strengths, but they were always meant to be engines of human invention and discovery rather than cubicles to constrain academic endeavors (Klein, 2010). To return to Einstein's analogy (see Chapter 1) that all disciplines of human knowledge are "branches from the same tree," the vitality of the whole depends on the strength of the foundation. The trunk of the tree represents the core from which disciplines draw in higher education—the centralizing force that directs students through a course of academic study. The branches—where Einstein located religion, arts, and sciences—can be seen both as the disciplines and as potential locations for integration. Branches grow away from the trunk, yet they remain integrally connected to the core strengths of the whole; they intersect and tangle in new ways as they grow. While the disciplines are powerful, they are not, and need not be, treated as fixed.

In his work on *Conceptual Foundations for Multidisciplinary Thinking* (1995), Stephen J. Kline explains the need to understand the connection between disciplines and the intellectual terrain as a whole:

> For at least a century, we have acted as if the uncollected major fragments of our knowledge, which we call disciplines, could by themselves give understanding of the emergent ideas that come from putting the concepts

> and results together. It is much as if we tried to understand and teach the geography of the 48 contiguous states of the United States by handing out maps of the 48 states, but never took the trouble to assemble a map of the country.

It is important to note that the purpose of this report is not to prescribe that institutions move away from disciplinary studies or, in the other extreme, to integrate all human knowledge within the educational experience of an individual student—that would be impossible—but rather to offer new insight into the impact of an approach to education that seeks to help students understand how the knowledge they have accumulated is connected.

THE DISCIPLINES DEFINED

Every field of study has its own epistemology that is learned through disciplinary preparation. The process of disciplinary education is characterized by certain conceptual gateways that are preconditions to any deep disciplinary understanding. As Meyer and Land (2003, p. 1) describe these conceptual understandings:

> A threshold concept can be considered as akin to a portal, opening up a new and previously inaccessible way of thinking about something. It represents a transformed way of understanding, or interpreting, or viewing something without which the learner cannot progress. As a consequence of comprehending a threshold concept there may thus be a transformed internal view of subject matter, subject landscape, or even world view. This transformation may be sudden or it may be protracted over a considerable period of time, with the transition to understanding proving troublesome. Such a transformed view or landscape may represent how people "think" in a particular discipline, or how they perceive, apprehend, or experience particular phenomena within that discipline (or more generally).

These threshold concepts point to the kinds of problems that each discipline is trying to solve or the contributions it is aimed at making to human understanding. However, these concepts tend to differ by disciplinary category and evolve over time as the "branches" of the disciplines further bifurcate or—in the case of established integrative discipline—intersect. Some integrative disciplines that are now relatively mature, such as Science, Technology, and Society; Gender Studies; Bioethics; and Computer–Human Interaction, historically have arisen at the intersections of existing fields. These new disciplines represent the potential for academic innovation through integration.

The Humanities

According to definitions adopted by the federal government, to study within the humanities, students focus on disciplines such as

> language, both modern and classical; linguistics; literature; history; jurisprudence; philosophy; archeology; comparative religion; ethics; the history, criticism, and theory of the arts; those aspects of the social sciences which have humanistic content and employ humanistic methods; and the study and application of the humanities to the human environment with particular attention to reflecting our diverse heritage, traditions, and history and to the relevance of the humanities to the current conditions of national life. (20 U.S.C. 952 (a))

A traditional liberal arts education included these humanistic disciplines as well as training in politics and abstract mathematics (Hirt, 2006; Lucas, 1994; Roche, 2013). Though classifications differ, the *qualitatively oriented* social sciences tend to be classified with the humanities.

The humanities teach close reading practices as an essential tool, an appreciation for context across time and space, qualitative analysis of social structures and relationships, the importance of perspective, the capacity for empathic understanding, analysis of the structure of an argument (or of the analysis itself), and study of phenomenology in the human world.

The Arts

The domain of the fine and performing arts

> includes, but is not limited to, music (instrumental and vocal), dance, drama, folk art, creative writing, architecture and allied fields, painting, sculpture, photography, graphic and craft arts, industrial design, costume and fashion design, motion pictures, television, radio, film, video, tape and sound recording, the arts related to the presentation, performance, execution, and exhibition of such major art forms, all those traditional arts practiced by the diverse peoples of this country, and the study and application of the arts to the human environment. (20 U.S.C. 952 (b))

The arts teach creative means of expression, understanding of different perspectives, an awareness of knowledge and emotions throughout the human experience, and the shaping and sharing of perceptions through artistic creation and practices in the expressive world. An art student's training in the methods and tools of a creative platform is complemented with studies in written and visual semiotics; critical and cultural theories and philosophies; historical antecedents that shape contemporary forms of cultural expression; and reflection-in-action through deep observation and constructive feedback.

The arts include not only all of these artifacts, intangible, tangible, and performative, but also the effect they have on people who participate and observe a given artistic expression. This impact has the capacity to build empathy and create new meaning for individuals in fields not limited to those traditionally associated with the arts, such as the social sciences.

The Sciences

The sciences include specialized fields covering the physical and mathematical sciences (i.e., chemistry, physics, and mathematics), the life sciences (e.g., cell biology, ecology, and genetics), the geosciences, computer science, and the quantitative social sciences (e.g., anthropology and sociology) (National Academies of Sciences, Engineering, and Medicine, 2006).

The sciences teach "the use of evidence to construct testable explanations and predictions of natural phenomena, as well as the knowledge generated through this process" (National Academy of Sciences, 2008). According to the UK Science Council, scientific methodology includes objective observation, evidence, experimentation, induction, repetition, critical analysis, verification, and testing (Science Council, 2017). The quantitatively oriented social sciences are generally included within the sciences. For example, the National Science Foundation, the federal agency whose stated mission is "to promote the progress of science; to advance the national health, prosperity, and welfare; [and] to secure the national defense,"[1] includes the social and behavioral sciences among its divisions and in its funding priorities.

Engineering

Engineering is the study and practice of designing artifacts and processes under the constraints of "the laws of nature or science" and constraints such as "time, money, available materials, ergonomics, environmental regulations, manufacturability, [and] reparability" (National Academy of Engineering and National Research Council, 2009, p. 17). It includes specialized engineering fields that focus on specific aspects of technology or the natural world, such as electrical, mechanical, chemical, civil, environmental, computer, biomedical, aerospace, and systems engineering.

Engineering teaches how to develop plans and directions for constructing artifacts and processes, such as computer chips, bridges, and drug

[1] For more information on the National Science Foundation see https://www.nsf.gov/about/glance.jsp (accessed July 16, 2017).

manufacturing processes (National Academy of Engineering and National Research Council, 2009). This is taught using design as a problem-solving approach that can "integrate various skills and types of thinking—analytical and synthetic things; detailed understanding and holistic understanding; planning and building; and implicit, procedural knowledge and explicit, declarative knowledge" (National Academy of Engineering and National Research Council, 2009, p. 37). The engineering design process is "generally iterative; thus each new version of the design is tested and then modified based on what has been learned up to that point" (National Academy of Engineering and National Research Council, 2009, p. 38). Engineering fields teach how to identify a need and design an efficient, functional, durable, sustainable, useful process or product that will meet that need.

Notably, the attributes of communication, teamwork, and ethical decision making (as well as the even broader attributes of critical thinking, applying knowledge in real-world settings, and lifelong learning) are increasingly considered core to the engineering disciplines, along with a greater acknowledgment of the responsibility of engineering to respond to human needs (e.g., The Engineering Grand Challenges,[2] National Academy of Engineering, 2008). This sea change dates roughly to the NAE's *Engineer of 2020: Visions of Engineering in the New Century* report (2004), which boldly articulated attributes of a twenty-first-century engineer, and the Accreditation Board for Engineering and Technology (ABET) "Engineering Criteria 2000," whose criteria significantly broadened the expectations for an engineering education. In response, large numbers of engineering programs have embedded teamwork experiences, communication, and ethics education into core engineering courses. In many instances, these new engineering programs have adopted an integrative model to achieve these learning outcomes.

Medicine

Medicine is the science or practice involved in "the maintenance of health as well as in the prevention, diagnosis, improvement, or treatment of physical and mental illness," and it includes the "knowledge, skills, and practices based on the theories, beliefs, and experiences indigenous to different cultures" (World Health Organization, n.d.). Medical fields aim to teach modern medical professionals five core competencies (Institute of Medicine, 2003, p. 45):

1. To provide patient-centered care (identify, respect, and care about patients' values, preferences, and needs; listen to, communicate with,

[2] For more information on the Engineering Grand Challenges, see http://www.engineeringchallenges.org/ (accessed July 16, 2017).

inform, and educate patients; share decision making and management with the patient; and advocate)
2. To work in interdisciplinary teams to cooperate, collaborate, communicate, and integrate care
3. To employ evidence-based practices by knowing where and how to find the best possible sources of evidence, formulating clear clinical questions, search for the relevant answers, and determine when and how to integrate these new findings into practice (evidence can include that which can be quantified, such as data from randomized controlled trials, laboratory experiments, clinical trials, epidemiological research, and outcomes research; evidence based on qualitative research; and evidence derived from the practice knowledge of experts, including inductive reasoning)
4. To apply quality improvement by understanding and measuring quality of care in terms of structure, process, and outcomes; assessing current practices and comparing them to relevant better practices; designing and testing interventions; identifying errors and hazards in care; improving one's own performance
5. To utilize informatics, such as using electronic data, communicating electronically, and understanding security protections

Medical fields teach how to analyze, conduct research on mechanics of the human body, examine relationships between bodies and environments, and make connections between disease and wellness.

These definitions highlight the unique aspects of each discipline and illustrate how the different disciplines consider and make use of different forms of evidence. Yet these definitions also demonstrate that the disciplines share the root purpose of creating knowledge for the betterment of humanity.

MULTIDISCIPLINARY, INTERDISCIPLINARY, AND TRANSDISCIPLINARY INTEGRATION

Integration can take multiple forms and can range from a relatively superficial intersection of disciplines to a deep integration of disciplinary knowledge. Often this range is characterized by the terms "multidisciplinary," "interdisciplinary," and "transdisciplinary" (Begg and Vaughan, 2011).

Multidisciplinary methods, typically considered the least integrative of the three, have been defined in several ways, yet converge on the idea that multidisciplinarity involves the process by which investigators from more than one discipline work from their disciplinary-specific bases to solve a common problem, either at the same time (Begg and Vaughan, 2011) or by sequentially applying ideas from multiple disciplines to the focal problem

(Hall et al., 2012). Multidisciplinarity, framed in this way, has been criticized as a temporary and often weak means of solving problems because of the superficial nature of that integration (Borrego and Newswander, 2010).

Through interdisciplinary approaches, in contrast, scholars work jointly from their disciplinary perspectives to address a common problem (Begg and Vaughan, 2011; Begg et al., 2015). The use of interdisciplinary methods requires team members to integrate their disciplinary perspectives—including concepts, theories, and methods—in order to solve the complex problem at hand (Hall et al., 2012).

Although interdisciplinary approaches are more integrated than multidisciplinary approaches, transdisciplinary research strategies require "not only the integration of discipline-specific approaches, but also the extension of these approaches to generate fundamentally new conceptual frameworks, hypotheses, theories, models, and methodological applications that transcend their disciplinary origins, with the aim of accelerating innovation and advances in scientific knowledge" (Hall et al., 2012, p. 416).

INTEGRATION IN THE CURRICULUM

As we have discovered in our review of integrative practices (and as will be apparent in Chapters 6 and 7), integration between STEMM fields, the humanities, and the arts can take many forms. Integration can be relatively brief in duration (e.g., a single assignment or unit within a course) or longer (e.g., a complete integrative course or a series of courses or related educational experiences). A wide variety of courses, programs, and other experiences can adopt an integrative approach, including first-year seminars, dual majors, minors, interdisciplinary courses and curricula, living-learning communities, and capstone projects. It can take place within a disciplinary or interdisciplinary major or within general education courses. It can reflect the world outside academia freed of academia's disciplinary silos. Integration can also be superficial and artificial when only one discipline is present for the learning design or delivery (Riley, 2015). Box 3-1 offers an example of a superficial integrative educational experience.

One definition of integration is that it merges contents and/or pedagogies traditionally occurring in one discipline with those in other disciplines in an effort to facilitate student learning. By this definition, integration can be as simple as using haikus or songs to convey scientific theories in class (Crowther, 2012; Pollack and Korol, 2013) or as complicated as developing full courses incorporating art, design, and engineering (Fantauzzacoffin et al., 2012; Gurnon et al., 2013). For example, exposure to the arts and humanities could demonstrate to students in STEMM fields the societal, economic, and political implications of scientific discovery and technological development (Grasso and Martinelli, 2010).

BOX 3-1
Is It Integrated?

Music in the Liberal Arts was an introductory course for undergraduates offered at a small liberal arts college to fulfill distribution requirements in the humanities. The goal of the course was to help students understand music from a variety of disciplinary perspectives. The organizer recruited faculty from many departments across the campus to explore with students in the course how their fields connected with music. For example, a psychology professor spoke about the emotional aspects of music and how music might influence mood and behavior. An anthropologist helped students understand music as a cultural phenomenon. A historian summarized how music changed over time, how it was shaped by events of the time, and how it influenced some of those events. A physicist explained the physics of sound waves, pitch, tone and tone quality, and amplitude. A biologist described the anatomy and physiology of the ear and hearing as well as the components of the human voice.

On the surface, this course appeared to epitomize an interdisciplinary approach to teaching and provide ample opportunity for the integration of concepts related to music. However, emerging research about human learning would suggest that this course's attempt to connect the disciplines was not, in fact, interdisciplinary, and that its effort to integrate conceptual knowledge was not likely to succeed. Although students were given opportunities to learn about music from experts in many fields, the faculty were never asked to discuss their presentations with one another or to attend each other's presentations. Thus it was highly unlikely that lecturers would deliberately or strategically tie together or explicitly reference the connections between the topics they discussed and those discussed by other professors. The course may have been *multi*disciplinary, but it was far from *inter*disciplinary.

Much more could have been done to help students make conceptual connections. For example, faculty colleagues from physics and biology could have worked together to pose related and connected problems for students to consider. To take a specific instance, the physics of sound explains that the longer a string is on an instrument, the lower its frequency will be. However, a bass singing voice is able to replicate the range of frequencies produced by cellos and many of the notes produced by a bass, even though human vocal cords are far shorter in length than any violin string. How can this seeming contradiction be reconciled? What are the biological adaptations that have taken place to enable this capacity in the human voice? What about in other species of animals? Such an approach could help students understand important physical and biological concepts and principles and also the *connections* among the disciplines of biology, physics, and engineering in the production of music.

Chapter 6 examines the evidence for positive outcomes from integrative learning experiences in three categories:

1. Through a single course (in-course integration), whether by offering students opportunities to observe a topic from multiple disciplinary perspectives, by creating a multidisciplinary teaching team, or by focusing on a theme that can be considered through various disciplinary lenses. In-course integration occurs when concepts and pedagogies from the arts and humanities are integrated into already established STEM courses, or vice versa, or when new interdisciplinary, multidisciplinary, or transdisciplinary courses are developed as part of a larger curriculum.
2. Through a combination of courses (within-curriculum integration), whether thematically linked general education courses, integrated elective courses, an integration of general education and majors, interdisciplinary majors or programs, or integrative seminars. Within-curriculum integration focuses either on adding non-discipline-related courses to a major curriculum or on developing an interdisciplinary, multidisciplinary, or transdisciplinary major with both arts and humanities and STEMM content.
3. Through extracurricular or co-curricular experiences, such as Maker Spaces and STEAM (science, technology, engineering, arts, and mathematics) clubs.

These three approaches share many similarities, and they overlap at times, but they tend to be structured differently and occur in different contexts. In the following sections we offer examples of courses and programs that fit within each of these categories. In Chapter 6, we discuss the known impact on students of many of these example courses and programs.

In-Course Integration

In-course integration occurs when concepts and pedagogies from the arts and humanities are integrated into already established STEMM courses, or vice versa, or when new interdisciplinary courses are developed as part of a larger, unintegrated curriculum. Box 3-2 describes examples of various forms of in-course integration. Examples include the Projects and Practices in Integrated Art and Engineering course taught at the Georgia Institute of Technology (Fantauzzacoffin et al., 2012), the use of digital video production to describe the fundamental process of neurotransmission in an introduction to neuroscience course at Emmanuel College (Jarvinen and Jarvinen, 2012), the Designing for Open Innovation course at The University of Oklahoma (Ifenthaler et al., 2015), and the Digital Sound

BOX 3-2
In-Course Integration in Practice

Imagine a professor teaching an introductory biology course for nonmajors, a course designed to help students become better-informed citizens with the knowledge to question and make educated decisions about scientific advances and policy. Because she wants to demonstrate to students that they are not alone in living in a time of change and uncertainty, she wants to introduce history, a humanities discipline, to her science course. To deepen her students' perspective on scientific change, she brings in a historian to help students understand the field of biology as it has developed over time. Perhaps the historian assigns a reading about the origins of epidemiology and John Snow's discovery in 1855 that cholera was transmitted through contaminated drinking water. This approach is a multidisciplinary version of STEMM–humanities integration. It incorporates a different disciplinary perspective (history) into its investigation but without leaving the boundaries of either discipline. Multidisciplinarity is additive, offering an additional set of content and insights that may be absent in the original discipline (Choi and Pak, 2006, p. 351).

Perhaps her history colleague will come to class on the day of the discussion or develop and help assess a student assignment in response to that discussion. In this way, the historian introduces not simply historical content but also the epistemology or analytical method of history. By contributing a historian's experience designing assignments, evaluating student work, and facilitating discussions, the history professor contributes historical content and the pedagogy of historical inquiry. The biologist might reciprocate in a history class, where the impact of scientific discovery on the course of human history could benefit from a scientist's expertise. For example, the biology professor might choose a reading and offer a mini-lecture on the specific scientific arguments made by Charles Darwin in On the Origin of Species to add nuance and specificity to the history professor's class session on nineteenth-century social and religious controversies.

Now imagine that the biology professor wants to offer a more ambitious integration. An interdisciplinary approach is more than additive; it synthesizes two separate disciplines to establish a new level of discourse and integration of knowledge (Klein, 1990). The field of epidemiology developed as a result of interdisciplinary integration, as John Snow learned the methods of social science and interviewed people in the London neighborhood affected by cholera and then plotted his findings on a map. Likewise, the biology professor might seek a partner from history or gender studies and take an interdisciplinary approach, transferring methods from one discipline to another.

Or suppose the biology and history professors wanted to develop a unit on public health. This unit might belong to a single course co-taught by the two faculty, be co-taught within two separate courses, or be part of an assignment contained in two thematically similar courses taught at the same time in order to combine some classes. If the faculty members work jointly from their specific disciplinary perspectives to teach a subject in an integrative way, addressing (for example) biological concepts and the implications of certain scientific policies or practices for women's lives, they are taking an interdisciplinary approach,

continued

BOX 3-2 Continued

one that "analyzes, synthesizes and harmonizes links between disciplines into a coordinated and coherent whole" (Choi and Pak. 2006). The two professors would model their own disciplinary training to make sense of the topic, and their students would learn the methods of two different disciplines as they explore the topic of public health. Functionally, this interdisciplinary offering might include assignments, discussions, and lectures that require a blended synthesis of content, techniques, and perspectives from different disciplines. The merging of biology and gender studies might lead to new questions, new perspectives, and an opening up of the field of "public health" to consider, for instance, the impact of domestic violence on the health of children (Fisher. 2011).

Finally, the biologist and gender studies professor might experiment with a transdisciplinary approach to integration. If multidisciplinarity is additive and interdisciplinarity is interactive, then transdisciplinarity is holistic in that the disciplines are subordinated to the overall system that includes the subject of inquiry; the disciplines might even disappear altogether, with students unsure where their professors are institutionally located. In transdisciplinary practice, the primary goal is not to convey the knowledge of any given discipline but rather to understand the world as it exists beyond the classroom, using whatever disciplinary tools are available and developing new knowledge, skills, and perspectives in the interplay among disciplines.

For example, the two faculty might offer a course in public health where students explore the problem of obesity. Here, the biologist might lead the students to understand the differences between innate biological mechanisms that promote obesity and the influence of external factors, while the gender studies professor might help the class explore the gender-specific disparities in obesity. After gaining familiarity with the many dimensions of this topic, the students are asked to

and Music online course developed by scholars at Wake Forest University (Shen et al., 2015).

Integration of the arts and humanities into STEMM courses, and the integration of STEMM into humanities and arts courses, can take many forms. Further, the goals and outcomes of integrative courses are diverse. Following here we offer a description of some existing efforts to integrate the arts, humanities, and STEMM fields, and vice versa, within the context of a single course and describe some of the goals of such efforts.

The integration of the arts and humanities into STEMM courses may inspire improved understanding of STEMM concepts, greater contextualization of STEMM subjects, new STEMM hypotheses and research questions, and enhanced innovation in STEMM. It may also support the development of twenty first–century skills in students, such as critical develop a problem statement. For example, they may discover that physicians' attitudes about obesity in women affect the likelihood of weight loss in female patients. They then set out to solve this problem using existing disciplinary tools as well as whatever new understanding they develop at the intersection of multiple disciplines and a complex social and biological problem. They might even enlist a dancer to bring in culturally relevant dances to a community of women, helping them develop habits of exercise through an art form. This is transdisciplinary integration. In transdisciplinary thought, disciplinary experts come together, working collaboratively to understand the problem and consider different approaches to solving it. In focusing on the problem to be solved rather than on their disciplinary norms, a transdisciplinary team is able to explore the limits and blind spots of their different disciplinary epistemologies, with each discipline undergoing modification to take on insights from the other fields. In this way, they create new knowledge, tools, and perspectives that differ from the foundation and approach offered by any one discipline.

All three approaches integrate STEMM and humanities disciplines, but they do so to different degrees and for different durations. These examples move from a single assignment, offering an easy way for faculty to experiment with integrative pedagogy, to more complex forms of integration. The later examples necessarily take more time to develop but are potentially more rewarding for students, faculty, and the institution itself. Consider the transdisciplinary case: as new knowledge is produced in the spaces between disciplines, it can give rise to new areas of study. Many established fields arose from these transdisciplinary interactions, including women's and gender studies, cultural studies, area studies, public and global health, robotics, and human–computer interaction, to name a few. Other fields are now emerging through the same process, including disability studies, product and game design, artificial intelligence, and art-technology programs such as entertainment technology (Graff, 2015).

thinking, communication skills, teamwork, and lifelong learning attitudes (see Chapter 6 for an expanded discussion). For example, the synthesis of mathematics and music has given rise to many courses, often using one of the topics to recruit students who may be reluctant to take a course on the other (e.g., a student who is comfortable learning about music but may be anxious about taking a mathematics course, or vice versa). Also, combining science, mathematics, and social justice can help both STEMM students and those from other disciplines appreciate the societal relevance of scientific and mathematical concepts and develop a critical eye for the use and misuse of evidence in public discourse (Chamany, 2006; Skubikowski et al., 2010; Suzuki, 2015; Watts and Guessous, 2006). As another example, the practice of origami has provided a nexus for artistic and mathematical energies, as evidenced by interdisciplinary symposia on many campuses, by the popularity of computer programmer-turned-origami artist Robert Lang as a guest speaker, and by the Guggenheim Award presented to MIT's Erik and Martin Demaine (Hull, 2006; Lang, 2012). Finally, some of the courses reviewed in Chapter 6 are associated with student outcomes that align with the twenty first–century skills, including critical thinking, teamwork, communication skills, and lifelong learning attitudes that employers are calling for and that will serve students in life and citizenship.

The integration of STEMM content and pedagogies into the courses of students pursuing a major or career in the arts and humanities can also take place in many different ways and for many different reasons. Among the courses we review in this report are those that strive to integrate STEMM with the goal of promoting greater scientific and technological literacy among humanities and arts majors, those that aim to harness STEMM tools to promote advances in artistic and humanistic scholarship and practice, and those that take STEMM, and the influence of STEMM on society, humanity, and nature, as the focus of humanistic and artistic inquiry.

Advocates for technological literacy among arts and humanities majors have created a variety of integrated courses, and a wide range of these have been surveyed and evaluated (Ebert-May et al., 2010; Krupczak, 2004; Krupczak and Ollis, 2006). In a 2007 workshop cohosted by the National Science Foundation and the National Academy of Engineering, John Krupczak and colleagues defined four main categories of efforts to foster "technological citizenship: survey courses, courses focused on a particular topic, design courses that involve students in technology creation, and technology in context" courses in which technology is critically connected to other disciplines. Longitudinal studies of technological literacy efforts have yielded a relatively robust set of technological literacy outcomes and methods for their assessment (NAE and NAS, 2006, 2012). See Chapter 6 for a review of the student outcomes associated with science and technology literacy courses.

Much as humanities and arts content often serve to contextualize STEMM content, some humanists have turned their lenses on technology, making STEMM the context for application of humanistic and artistic methodologies. The interdisciplinary discussions fostered by the Society for Literature, Science, and the Arts in its journal *Configurations* served as a forum for such scholars as Katherine Hayles and Donna Haraway to discuss what it means to be human in a "posthuman" world (Gerrans and Hayles, 1999) or an increasingly technophilic world (Haraway, 1994). In Chapter 4, we describe undergraduate courses that use STEMM subjects as topics for humanistic and artistic inquiry.

Also, some humanists and artists see integration with STEMM fields as necessary for addressing the challenges of our century. Through the Humanities Connections grant program, the National Endowment for the Humanities (NEH) has funded curricular development that integrates humanistic study with other disciplines, either in general education or in major fields of study, to advance this goal. Former NEH chairman William D. Adams explained the goals of the Humanities Connections program:

> The most important challenges and opportunities of the 21st century require the habits of mind and forms of knowledge fostered by study of the humanities. The NEH Humanities Connections grant program will help prepare students in all academic fields for their roles as engaged citizens and productive professionals in a rapidly changing and interdependent world.[3]

The NEH particularly encourages projects that foster collaboration between humanities faculty and their counterparts in social and natural sciences and in preprofessional programs in business, engineering, health sciences, law, and computer science. The inaugural grant competition funded 18 programs in medical humanities, environmental humanities, urban humanities, creative and ethical entrepreneurship, ecoliteracy, digital humanities, humanities and engineering, and other integrative topics.

Further, some humanists claim that STEMM pedagogies can strengthen learning in humanities courses. For example, Cavanaugh (2010) argues that humanists should use approaches borrowed from the cognitive sciences, such as problem-based learning, wikis, service learning, and other software tools, to boost the outcomes associated with the humanities. Such approaches can result in outcomes that include comfort with ambiguity, problem-solving abilities, more astute questioning, and drawing relationships through the use of metaphors, similes, and demonstrations.

Although humanities and art scholars always have used technical tools in their research and pedagogy, more and more scholars and students are engaging with the sophisticated technical tools grouped under umbrella terms such as “digital humanities” and “big data.” Examples include geographic information system (GIS) mapping (Bodenhamer et al., 2010), the use of databases for research, and rapid prototyping or 3D printers. Box 3-3 describes the use of engineering design as one such integrative tool.

Within-Curriculum Integration

Within-curriculum integration focuses either on adding non-discipline-related courses to a major curriculum or developing an inter- or transdisciplinary major with both arts and humanities and STEMM courses. Examples of strong programs include Sixth College at the University of

[3] For more information on the National Endowment for Humanities’ Connections Grant Program, see https://www.neh.gov/news/press-release/2016-04-29.

BOX 3-3
Engineering Design as an Integrative Tool

The engineering design process, which synthesizes humanistic, social, creative, and analytical skills, is one avenue for meaningful integration of a range of disciplinary methods and values in courses. Frameworks for the engineering design process use varying nomenclature to describe the same essential elements: need finding (or empathy), problem definition and framing, creative idea generation, prototyping, and testing and analysis. The process is iterative, and communication with multiple stakeholders is critical throughout the process. While this is an engineering methodology, it shares with the arts an emphasis on creativity, and with the humanities and social sciences a comfort with the ambiguity of nonunique, context-specific solutions.

The overlap between prototyping and making means that makerspaces and design studios are often housed in engineering spaces, but these activities, like design itself, are not limited to engineering students. In fact, making is also a studio art and an act of creation—what is "designed" might be a story, a textile, or a 3D-printed widget. In critical making, students apply analytical faculties from the humanities and social sciences to this creative endeavor (Somerson, 2013). Crawford has persuasively contended that such handwork is also "soul craft," enriching students' humanity and personhood as well as their professional development (Crawford, 2009).

The importance of effective communication and collaboration with fellow designers integrates additional elements in the design process. Interpersonal dynamics and written and oral communication are critical to effective and successful design. Through project-based collaboration, students develop both skills and confidence in the value of their own expertise (Kelley and Kelley, 2013). The best examples of such collaborations are ones in which all members bring distinct skills and disciplinary perspectives to bear on shared goals rather than ones in which humanities and arts students simply support an essentially technical design challenge.

California, San Diego (Ghanbari, 2015), the Connections program for first-year engineering students at the Colorado School of Mines (Olds and Miller, 2004), and the integrated program for first- and second-year students implemented by the College of Engineering at Texas A&M University (Everett et al., 2000; Malavé and Watson, 2000).

The engineering design process (described in Box 3-3), can also provide an organizing principle for within-curriculum integration. For example, the Massachusetts Institute of Technology's Terrascope is a first-year program that supplements introductory courses with problem-based experiences and cross-disciplinary teams (Lipson et al., 2007). The iFoundry program at the University of Illinois began as an infusion of philosophical and other perspectives into engineering education and is now a multifaceted "cross-disciplinary curriculum incubator" for project-based learning, entrepreneurship and innovation experiences, and methods for enhancing students' intrinsic motivation. An initiative at Smith College has involved faculty, students, and staff from all disciplines in a design thinking community to reimagine the liberal arts. This project embraces "radical collaboration to encourage the unconventional mixing of ideas, thereby creating a culture where ideas (and the technologies that help us realize these ideas) belong simultaneously to no one and everyone" (Mikic, 2014).

Many universities with strong STEMM and liberal arts programs have a long history of offering programs in science, technology, and society (STS), also sometimes called "science studies." Generally, these programs apply the methods and values of humanities and social science inquiry to the natural sciences and engineering. They teach students to understand and critique science and technology in their historical, political, and cultural contexts and to appreciate the social forces that surround and shape advances in scientific knowledge and technology (Akcay and Akcay, 2015; Han and Jeong, 2014). In these programs, students must understand the nature of scientific and technical inquiry and innovation as well as develop the critical thinking skills associated with political science, history, sociology, anthropology, and ethics. Each program tends to occupy a particular niche, both in the broader field of STS and at its own institution. For example, the program at the University of Virginia is housed within the engineering school and offers courses such as engineering ethics to engineering undergraduates. Others, for example, programs at Lehigh University and Virginia Tech, are housed in colleges of arts and sciences and were founded with the vision of attracting both engineering and liberal arts students. The growth of STS has helped demonstrate the many ways in which science depends on technological advances, as well as the dependence of both science and technology on economic, social, political, and cultural factors.

Another curricular program that integrates humanistic and STEMM fields centers on the profound ethical questions resulting from rapid scientific and technological advances in medicine. Bioethics is now a well-established integrative discipline in which students develop the tools and context for moral discernment in life sciences, medicine, and biotechnology, infusing their analyses with content and perspectives from law, policy, and philosophy (Leppa and Terry, 2004; Lewin et al., 2004; Vaughn, 2012). More broadly, ethics is a standard (and often required) component of research programs in the sciences (NAS, NAE, and IOM, 2009). Premed and engineering curricula, because they are more oriented toward professional tracks, also provide opportunities to integrate philosophical, sociological, and humanistic modes of inquiry and content as part of ethics instruction.

The Grand Challenges issued by the National Academy of Engineering in 2008 are motivating engineering educators and practicing engineers to consider problems that are inherently sociotechnical and are intertwined with geopolitical, economic, philosophical, and cultural factors. Institutions that develop Grand Challenges project experiences recruit students from many majors. In working together to define design problems and to identify context-specific issues and possible solutions, students from all backgrounds gain appreciation for the methods, values, and history of other disciplines. When designed to explicitly include nonengineering students, the aim is for students to develop a mutual literacy in each other's disciplines and collaborate in this shared space (National Academy of Engineering, 2012).

Worcester Polytechnic Institute's (WPI) Great Problems Seminars address a wide range of vexing global sociotechnical problems, including the Grand Challenges (Savilonis et al., 2010). Since 2007, this team-taught problem-based learning course has engaged first-year students in "interdisciplinary, not multidisciplinary" discussions and design projects related to global concerns. Faculty teams are multidisciplinary, pairing, for example, a chemist with an economist. WPI has used both internal and external assessments to refine course outcomes, structure, and delivery. Faculty members have also developed a handbook to enable additional WPI faculty to join the Great Problems teaching team and to disseminate effective strategies. Preliminary assessment data suggest that students in this program showed evidence of teamwork, empathy, and integrative learning (DiBiasio, et al., 2017).

It is important to note that within-curriculum integration can take many different forms. One way that within-curriculum integration often occurs is through general education programs (see Box 3-4). A common form of general education in colleges and universities is the "cafeteria approach," where students take a selection of different courses outside their major and are thereby considered to be generally educated. Schools in the University of California system employ a more organized approach, where students take classes according to prescribed thematic clusters. The University of California–Merced, in particular, has launched an innovative first-year undergraduate course called "Core 1: The World at Home." Core 1 introduces students to the range of scholarly inquiry at the university, all in the span of a one-semester, writing-intensive, integrated curriculum that encourages them to make their own connections among the disciplines while practicing both qualitative and quantitative analysis. The course entails a series of 15 weekly 1-hour lectures (given by different faculty from across the disciplines) whose subjects students process in 2.5 hours of small-group discussion sections (the instructors of which assign and grade all course work) and a coordinated, cumulative sequence of written assignments (Hothem, 2013).

BOX 3-4
Integrated General Education Program at the University of Virginia

The University of Virginia has three different programs in general education: the Traditional Curriculum, the New Curriculum, and the Forum Curriculum. The traditional Curriculum follows the common distribution requirements and would be most useful for transfer students, while the New Curriculum is closely related to a liberal arts program and requires courses in three areas: engagements, literacies, and disciplines. The New Curriculum also requires freshmen to take four seven-week engagement courses.

The third option at the University of Virginia is the Forum Curriculum, which has similarities in integration to the general education programs at LaGuardia Community College and Guttmann Community College (both City University of New York); Portland State University; and selected California State University campuses). The Forum Curriculum takes an integrated approach and offers students general education under seven different themes. The current themes that have been offered are creative processes and practices; epidemics; human influence on the environment; mobility and community; visions of the good; food, society, and sustainability; and space, knowledge, and power. Each theme has two faculty members who provide oversight, a required freshmen seminar, a capstone course, and clusters of courses that generally include requirements in the humanities, social sciences, and natural sciences that are related to the particular general education theme. This theme-based curriculum allows students to become deeply engaged in areas that may be outside the major and provides integration across multiple disciplines with courses related to the theme. Moreover, the requirement of the freshmen seminar and a capstone course within the theme gives additional structure toward an integrated learning experience focused on a theme that is outside students' majors.

WPI's 47-year-old curriculum offers another example of integration in the context of general education. This curriculum, known as the WPI Plan,[4] brings vertical integration through general education requirements to every undergraduate. Since 1970, every WPI undergraduate has completed three general education projects in addition to the (optional, but popular) six-credit Great Problems Seminar described earlier. Two of these three required projects are deeply integrative. Under the WPI Plan, all undergraduates complete an 18-credit Humanities and Arts Project allowing them to pursue a creative or scholarly project of their choice through the lens of the humanities or arts. During the junior year, students complete a nine-credit

[4] For more information on the WPI Plan, see https://www.wpi.edu/project-based-learning/wpi-plan.

Interactive Qualifying Project, an open-ended interdisciplinary team project addressing some topic at the intersection of technology and human need. In the senior year, they complete a nine-credit Major Qualifying Project, the equivalent of a senior thesis or research project. As these projects take place at every year of the undergraduate course of study, students interweave integrated, purposive projects with coursework in their major. See Chapter 6 for additional examples of integrative general education programs.

In addition to integrative general education, global education can offer opportunities for building integrative competencies. For example, the University of Rhode Island's successful International Engineering Program (IEP), in which engineering students double major in a foreign language and an engineering discipline (coupled with a study-abroad experience), has grown steadily and expanded to several language tracks. The IEP has produced other, less-anticipated benefits: "Women have enrolled in engineering in increasing numbers . . . and the academic quality of Rhode Island's engineering students has improved" (Fischer, 2012). Although such programs are built to couple STEMM with language ability, their appeal to students suggests that integrative projects that focus on the Grand Challenges in a global context may strengthen not only all students' global citizenship but also the perceived, real-world relevance of the contributions of both STEMM fields and the arts and humanities. Blue et al. (2013), Nieusma (2011), and others have documented the challenges and rewards of such global projects for a wide range of students.

Integration can also take place in the context of learning communities. One explicitly arts and STEM integrative learning community is the University of Michigan's Living Arts program,[5] themed around the creative process and funded by the provost's office and academic units that self-identify as "maker" units. These include the School of Music, Theatre and Dance; Stamps School of Art and Design; Taubman College of Architecture and Urban Planning; and the College of Engineering. They have a required first-semester four-credit curricular course, Introduction to Creative Process, co-taught by one instructor from each of the four academic units, plus a writing instructor. Students receive academic credit fulfilling the university's first-year writing requirement, as well as experience hands-on making through the lens of creative process exploration within all four disciplines. At the graduate level, the University of Michigan also has one of the only engineered interdisciplinary graduate residences, the Munger Graduate Residences. It actively recruits students from all 19 academic units on campus, and places them in living suites based on interests, not disciplines. Their website states, "Experience true multi-disciplinary collaboration. The

[5] For more information on the University of Michigan's Living Arts program, see https://livingarts.engin.umich.edu/about/.

world increasingly presents challenges that cut across multiple disciplines and skillsets. At the Munger Graduate Residences, a diverse mix of graduate and professional students from various fields live, study and interact together, building a culture of collaboration."[6]

Learning communities are common at community colleges, as they are considered a high-impact practice (Keup, 2013; Tinto, 2003). Of the examples of integrative programs this committee considered at community colleges, the vast majority took place in the context of a learning community. See Box 3-5 for one example of a successful integrative program at Guttmann Community College. Other notable programs are in place at LaGuardia Community College,[7] Maricopa Community College,[8] and Seattle Central Community College.[9] Given that they generally have limited resources and a relatively short time with students, community colleges have had to be particularly innovative in producing integrated approaches to general education—for example, combining training in mathematics, writing, historical analysis, and natural sciences in single courses. According to administrators at these institutions, one of the most difficult challenges they face is policy makers' misconceptions that liberal education and vocational training are unrelated and that the former offers no added value to the latter.

Co-curricular and Extra-curricular Integration

Co-curricular and extra-curricular integrative opportunities include internships, faculty-run labs and makerspaces, and interdisciplinary research programs. Some of the most popular programs include the Maharam STEAM Fellows at the Rhode Island School of Design (Rhode Island School of Design, 2016), the Launch Lab at Youngstown State University (Wallace et al., 2010), the Institute of Design at Stanford University (Borrego et al., 2009), and the "Dance Your PhD" competition hosted by *Science* magazine (Bohannon, 2016; Shen et al., 2015).

The movement associated with the acronym STEAM provides many examples of co-curricular integrative initiatives. Official STEAM student clubs have expanded to many campuses, including Brown, MIT, and Harvard, as students focus on "uniting the Arts with STEM" to "ignite

---

[6] For more information on the University of Michigan Munger Graduate Residencies, see http://mungerresidences.org.

[7] For more information on Learning Communities at LaGuardia Community College, see https://www.laguardia.edu/ctl/Learning_Communities.aspx.

[8] For more information on Learning Communities at Maricopa Community Colleges, see https://hr.maricopa.edu/professional-development/learning-communities.

[9] For more information on Learning Communities at Seattle Central College, see https://seattlecentral.edu/programs/college-transfer/learning-options/learning-communities.

BOX 3-5
Integration at a Community College

The Stella and Charles Guttmann Community College of the City University of New York uses an integrated, interdisciplinary approach to increase student retention, understanding, and interest. Founded as the New Community College (NCC), it has an educational framework that includes mandatory full-time enrollment for first-year students, student participation in a common first-year experience, a limited choice of majors and electives, a required capstone course, student services such as mentoring and advising, comprehensive and continuous assessment, a well-coordinated and student-centered admissions process, a mandatory summer bridge program, co- and extracurricular activities, the formation of a learning community, and a focus on research in writing-intensive courses. All programs are built around the idea of using New York City to enable students to develop connections between their work and their environment.

One of the most important elements of the program at NCC is student buy-in. The integrated instructional model is explained to prospective students, and all agree to take part in NCC's experimental interdisciplinary curriculum. Once students enroll, they are assigned to a cohort during the summer bridge program and remain in this group until they choose a major. NCC offers a curriculum that does not require separate disciplinary courses so that students can build developmental skills, which many of them need, while still learning college-level material. Integration of multiple subjects promotes application of knowledge beyond mere memorization and retention.

The first year at NCC is structured around an integrated course known as the City Seminar. Learning goals center around a single issue per semester that is relevant to students' lives and experiences, such as sustainability or immigration. The first semester is built around four components: critical issue, quantitative reasoning, reading and writing, and group work space. In critical issue, students

communications between disparate fields in academia, business, and thought" (STEAM, 2016, para. 1). STEAM efforts have gained legislative support through House Resolution 319, introduced in 2012 and still under committee consideration, which "expresses the sense of the House of Representatives that adding art and design into federal programs that target Science, Technology, Engineering and Math (STEM) fields, encourages innovation and economic growth in the United States."[10] Notable STEAM efforts include instruction in hand drawing (at the University Illinois) and narrative and role playing (at the University of Delaware).

[10] H.Res.319—Expressing the sense of the House of Representatives that adding art and design into federal programs that target the science, technology, engineering, and mathematics (STEM) fields encourages innovation and economic growth in the United States. 112th Congress (2011–2012).

learn to develop their critical thinking skills and to examine issues from multiple perspectives. In quantitative reasoning, they learn to analyze numerical aspects of real-life situations. In reading and writing, they build a deeper understanding of content through intertextual connections and reflective writing. In group workspace, they gain an understanding of their learning processes and have the opportunity to develop skills through project-based and experiential activities. Each component is taught by a different instructor, and these four faculty members make up the instructional team for a cohort. In the second semester of City Seminar, the reading and writing component is replaced by English Composition I.

Student and faculty collaboration is vital to NCC's success. Faculty work together to create "signature assignments" that seamlessly integrate all major components of the program, with 11 defined learning outcomes that cover a wide range of skills and disciplines. The curriculum focuses on "skill spines," faculty-defined topics that should be covered in every offering of the City Seminar. Once the theme of each seminar has been decided, faculty identify texts and resources that can be used to address each component. This material is compiled in a City Seminar I Instructional Binder provided to all faculty teaching the course.

The City Seminar is still evolving to meet student and teacher needs, but initial results are promising. Two years after entering, 27.0 percent of the students in NCC's inaugural class graduated with their associate's degree, compared with a City University of New York–wide 2-year graduation rate of 4.1 percent. More broadly, the program has revealed how effective communication among faculty and students of diverse backgrounds and interests allows for a broader appreciation of multiple points of view and the ability to engage in open discussions. Though not all students can adhere to the college's requirements, such as attending full time, the program has demonstrated positive results, and other schools have adopted similar models.

SOURCE: Saint-Louis et al., 2015.

Integrative, experiential learning experiences offer students an opportunity to appreciate both their own and others' contributions to a shared outcome. Such projects may be commercially or socially entrepreneurial, community based, concerned with social justice, or focused on a combination of valued goals, thereby developing each team member's skills and perspectives in service of a larger goal. They may employ various pedagogical tools, such as problem-based learning, design thinking, or other collaborative processes.

A platform for experiential learning experiences can often be found in campus-based centers for innovation, creativity, and/or entrepreneurship. In nonprofit and public-sector projects, social innovation can be as relevant as innovative commercial ventures (Gulbrandsen and Aanstad, 2015). These kinds of experiences can strengthen both STEMM and arts and humanities students' abilities to value the merits of their own disciplinary training while learning more about the contributions of others (Brown and Kuratko, 2015).

STUDIES OF INTEGRATIVE EXPERIENCES DO NOT ALWAYS INVOLVE INTEGRATION OF THE HUMANITIES, ARTS, AND STEMM

Although this study uses the term "integration" to refer specifically to the integration of the humanities and arts with STEMM fields, higher education scholars consider integrative educational experiences more broadly. Specifically, some scholars view "integration" as a learning outcome unto itself (i.e., "integrative learning") and characterize it as a process or mechanism that helps students integrate or bring together ideas as an embedded element of a curricular or co-curricular program or initiative. Importantly, this may occur in the context of the integration of the humanities and arts with STEMM fields, or it may occur in other educational contexts. Indeed, scholars who work with the National Survey of Student Engagement have established that the integrative experience may involve any college process or mechanism that students identify as helping them integrate or bring together ideas and may not be something that is bound to any given curricular or co-curricular context (Laird et al., 2005). Though we focus exclusively on the integration of the humanities, arts, and STEMM subjects in this report, we offer this description of the larger context in which scholars have considered integration to acknowledge that the type of integration this study is dealing with falls within a larger body of research in higher education.

WHAT IS INTEGRATIVE LEARNING?

In the higher education research literature, the term "integration" can refer both to the design of a learning experience (e.g., a course that integrates medicine and the arts) and to a student's cognitive experience that unifies different disciplinary approaches (e.g., an assignment that asks students to integrate engineering design principles into an ethical decision-making scenario). We offer here two published definitions of integrative learning that provide some insight into the anticipated student outcomes of an integrative learning experience: one from higher education researcher James Barber, and a second from the widely used and often lauded Association of American Colleges and Universities (AAC&U) rubric for integrative learning.

Barber (2012, p. 593) defines integrative learning as

> the demonstrated ability to connect, apply, and/or synthesize information coherently from disparate contexts and perspectives, and make use of these new insights in multiple contexts. This includes the ability to connect the domain of ideas and philosophies to the everyday experience, from one field of study or discipline to another, from the past to the present, between campus and community life, from one part to the whole, from the abstract to the concrete, among multiple identity roles—and vice versa.

Extending this idea, the AAC&U has developed an assessment rubric designed to help educators ascertain when integration is occurring in their students' work (AAC&U, 2010). From the association's perspective, students are demonstrating integration as a learning outcome when they are able to

- connect relevant experiences and academic knowledge;
- see and make connections across disciplines and perspectives;
- adapt and apply skills, abilities, theories, or methodologies gained in one situation to new situations;
- communicate in language that demonstrates cross-disciplinary fluency; and
- demonstrate a developing sense of self as a learner, building on prior experiences to respond to new and challenging contexts.

These capabilities point toward a distinctive form of learning (Barber, 2012). The definitions from Barber and AAC&U demonstrate a growing commitment to understand not only experiences that are integrated but also how those experiences might contribute distinctively to student learning. Perhaps exposing students to integrated learning experiences will not only promote existing learning and career outcomes but also spur a distinctive form of learning not captured by or part of other learning dimensions (e.g., cognitive development, critical thinking, pluralism, etc.). For an excellent discussion of this issue, see Barber (2012) and Youngerman (2017). But whether the integration of the humanities, arts, and STEMM disciplines leads to integrative learning remains an open question. As Chapter 5 demonstrates, the available research does not speak directly to this question. The committee would urge future research to consider this question as, hypothetically, certain approaches to the integration of the humanities, arts, and STEMM should promote integrative learning.

Assessment of integrative learning is important for understanding the student experience in a course or program that integrates the humanities, arts, and STEMM disciplines. Unless students are deliberately making connections across disciplinary domains, integration may not be taking place despite the programmatic design choices of educators. Although integrative programs and initiatives have been studied for their relationship to learning outcomes (see Chapter 6), very few scholars have examined what students are integrating or how participation in these programs and initiatives helps students with the integrative process (i.e., how these students are bringing together information). Rather, scholars assume that integration is occurring as a result of students' participation in these programs and initiatives and suggest that associations between participation and learning outcomes are based on the assumed integration occurring. The next chapter further explores the challenges of assessing learning outcomes in higher education, in general, and the challenges of assessing the impact of programs and courses that integrate the humanities, arts, and STEMM, specifically.

4

The Challenges of Assessing the Impact of Integration in Higher Education on Students

How can we know the impact of an educational experience on a student? To answer this question, we first must ask what kinds of student outcomes we expect from that educational experience. If the answer is "a better SAT score," "a better graduation rate," or "more students liked the class," then we can provide a quantitative measure of the impact. Indeed, many national education initiatives in the past two decades have set such quantitative benchmarks as proxies by which schools can measure their effectiveness in promoting student learning.[1] But if the answer is "a change in perspective" or "new insights into the human condition," narrative and qualitative evidence would likely offer more information.

As the committee approached its task to examine "the evidence behind the assertion that educational programs that mutually integrate learning experiences in the humanities and arts with science, technology, engineering, mathematics, and medicine (STEMM) lead to improved educational and career outcomes for undergraduate and graduate students," committee members found it necessary to first examine the nature and meaning of "evidence," and how different kinds of evidence (e.g., qualitative, quantitative, narrative, anecdotal observation, etc.) inform decision making in real-world contexts. This examination of the meaning and nature of evidence revealed that different stakeholders have different perceptions of what forms of evidence are appropriate and informative in decision making.

---

[1] Goals 2000: Educate America Act, Pub. L. No. 103-227 (1994). No Child Left Behind Act of 2001, Pub. L. No. 107-110, § 115, Stat. 1425 (2002). America Competes Act of 2007, Pub. L. No. 110-69 (2007).

We found that, just as there are those who believe that the only legitimate approach to evidence-based decision making is one that relies on quantitative, controlled, randomized studies, there are many at the opposite end of the spectrum who believe that quantitative approaches are insufficient for evaluating deeply human and social issues because they reduce the complex experiences and behaviors of individuals to numbers and the characteristics of a population. The committee concluded that these perspectives represent the extreme ends of a continuum and, in most cases, evaluating something as complicated as human learning will require an approach that lies somewhere between strictly qualitative or quantitative methods. The optimal approach to evaluation will vary according to the questions that researchers or course designers seek to answer and the real-world constraints that influence which methodological approaches are possible.

This chapter describes the committee's consideration of the value and nature of different forms of evidence and the realities of evidence-based decision making in real-world contexts. We discuss the challenges of generalizing the "evidence of improved educational and career outcomes" when different stakeholders have different interpretations of positive educational outcomes, assess outcomes in different ways, and use different pedagogical structures to approach integrative teaching and learning. We conclude that it is appropriate and necessary to consider multiple forms of evidence when considering the impact of an educational experience on a student, and that approaches to evaluating the impact of courses and programs that integrate the humanities, arts, and STEMM will necessarily be diverse and should be aligned with the specific learning goals of the course or program in its own institutional context.

EVIDENCE-BASED DECISION MAKING IN REAL-WORLD CONTEXTS

As the committee considered the value of multiple forms of evidence in decision making, we found it helpful to consider examples of evidence-based decision making in real-world contexts. A brief examination of real-world, evidence-based decision making demonstrated that, although we often demand longitudinal, controlled, randomized, causal studies, it is not always possible, or even necessary, for researchers to collect this form of evidence for decision making. This is true even when we consider how evidence is gathered in fields traditionally associated with rigorous quantitative methods.

Take the field of medicine, for example. If we study the process by which drugs are brought to market, from their initial discovery to their final approval by the U.S. Food and Drug Administration (FDA), we learn that not every drug is required to be tested using double-blinded, randomized patient populations with a placebo control.[2] In fact, many of the drugs currently on the market, including morphine, penicillin, vitamins, and aspirin, have never undergone FDA testing.[3] Medical professionals consider the combination of efficacy and safety of certain compounds so obvious that they have been "grandfathered" into use. Other drug trials cannot adhere to strict experimental conditions because of the nature of certain diseases and conditions. For instance, some chemotherapies cannot be blinded because severe side effects make patients receiving the treatment easily distinguishable from those who received the control.[4] For diseases that are extremely rare, there may be too few patients for control groups, so cross-over studies—in which a single group of patients is alternately treated and then taken off the treatment several times—are sometimes used instead (Delaney and Suissa, 2009).

Importantly, in many instances, clinical trials are performed only after preliminary studies have indicated the probability that clinical trials would succeed. Furthermore, many clinical trials in medicine begin only after an "n of 1" observation suggests to a biomedical researcher that a unique, unexpected, and possibly useful phenomenon has revealed itself. An example is John Fewster's observation that a farmer who had been infected with cowpox was subsequently protected against smallpox (an example of anecdotal evidence in the form of clinical observation) (Boylston, 2018). In such cases, the n of 1 observation is followed up with further observations that may confirm or disconfirm the initial hypothesis. These observations are followed by experiments to determine whether the phenomenon can be reproduced—often without controls or measurements—in an animal or human subject under laboratory conditions. If such uncontrolled experiments are successful, fully controlled experiments may be performed in animals. If these warrant further study, then a safety trial, without any measurements of efficacy, will be done on a very small cohort of healthy human subjects who cannot benefit from the treatment. If the treatment is safe, then another very small clinical trial will be done on patients with end-stage disease to determine whether the treatment may be efficacious. Only after all these stages of preparation have been completed is a "gold-standard" study (double-blind, randomized, placebo-controlled) performed.

This description of the generation and use of multiple forms of evidence in medicine makes several relevant points for evaluating the impact that integrating the arts and humanities with STEMM has on students in

---

[2] See https://www.fda.gov/downloads/drugs/guidancecomplianceregulatoryinformation/guidances/ucm073137.pdf (accessed August 17, 2017).

[3] See https://www.fda.gov/Food/IngredientsPackagingLabeling/GRAS/ (accessed August 17, 2017).

[4] See https://www.cancer.gov/about-cancer/treatment/research/placebo-clinical-trials (accessed August 17, 2017).

higher education. The first is that evidence, regardless of type, is developed in stages. New discoveries begin with observations that then progress to more formal study. Although it is true that an anecdote offers limited predictive evidence, it is also true that an anecdote may be the first step toward a meaningful discovery. Collecting evidence on the impact of anything, be it a drug or a curricular intervention, is a process that proceeds from uncontrolled observations or interventions through to formal qualitative and/or quantitative analyses, which may or may not eventually lead to randomized, controlled trials. To accept only the end point as legitimate evidence would undermine the process by which evidence is gathered and informs decision making. The committee therefore cautions against a one-size-fits-all decision-making process or framework, and instead encourages evaluative practices that incorporate multiple forms of evidence, along with contextualized metrics of quality and value.

Not all situations lend themselves to randomized, controlled trials. Integration seeks to achieve a wide variety of outcomes—content mastery, confidence, empathy, creativity, communication skills, teamwork, critical thinking, motivation, life-long learning attitudes, among others—and not all are equally amenable to quantitative approaches. Even when a quantitative approach would be most informative, it is often the case that a controlled, longitudinal, blinded, randomized study is not possible given the circumstances. In research on education there are limits to the number of variables that we can control, and even when we have considerable control over certain variables, there are some aspects of student experience that we cannot control—some students come to class every morning having eaten breakfast, while others do not; some students have parents that encourage them to study, while others do not; some students are holding down a job and supporting a family, while others do not have these responsibilities, and so on. Further, randomly assigning students to curricular "treatment groups" in higher education research is often challenging because course taking, program participation, and major selection are all areas where students are given the agency to choose to be involved or not. Even when choosing to be involved in something, students often vary in their level or manner of involvement. Though there are some methods to deal with nonrandom assignment, accounting for all the possible variations is an extremely difficult task.

The committee expects that integrated educational experiences will have multiple impacts on students, some of which lend themselves to quantitative or qualitative approaches to data collection and analysis, whereas others defy traditional measures of impact. For example, can the impact of a work of art or a musical performance always be sufficiently described in words or numbers? One could imagine a host of qualitative and quantitative data we could collect in an effort to try to address such a question, but essential elements of a transcendent musical experience would still be overlooked in any conventional analysis. We maintain that a combination of evidence will be necessary to demonstrate the effects of integration on students, and such a combination will be more convincing than any one type of evidence. The collection and analysis of evidence regarding integration will be ongoing, cumulative, and multifaceted.

THE CHALLENGES AND LIMITATIONS OF RESEARCH ON INTEGRATION

There are unique challenges to the evaluation of integrative educational courses and programs that stem from the fact that the various disciplines generate, evaluate, and disseminate evidence in different ways. Scholarly evidence in the fine arts is different from scholarly evidence in the humanities, and both are distinct from scholarly evidence produced in STEMM fields. Indeed, differences in judgments about what counts as evidence to warrant particular sorts of claims are among the key elements that define and distinguish disciplines. Communities of scholarly practice use different epistemologies and different methods of research (Kuhn, 2012). The artifacts they produce have different purposes and audiences. They evaluate the relevance and quality of evidence differently and consider different types and standards of acceptable evidence in making judgments. This is one of the greatest strengths of interdisciplinary work to begin with: the ability and willingness to draw from these rich traditions of data, research, and analysis.

Similarly, integrative teaching and learning can also disrupt the conventions for judging quality. Within particular disciplines, scholars seek evidence of the quality of learning that is valued within that discipline. For instance, while all students might be expected to give nearly identical answers when solving a mathematical equation, an instructor might expect quite different interpretations of the same passage from *Hamlet*. Yet for both subjects there are better and worse, correct and incorrect answers. The desired and measurable learning outcome in the math course might be the ability to fundamentally understand and logically solve the equation, while in the English course it may be the capacity to engage in a close reading and informed interpretation of a literary text. Understanding the impact of a course that integrates mathematics and English could be confounded by the challenge of developing an assessment tool that can adequately compare two very different approaches to education or discipline-specific conceptions of desirable learning outcomes. This is often further complicated by a lack of appropriate baseline data by which to measure any change in outcome with the introduction of a new pedagogical or curricular approach. The evidence gathered may or may not appear valuable, and evidence may be applied, understood, or interpreted in diverse ways. Thus judging the quality of scholarly productions that cross disciplines and fields almost invariably involves different conventions for making judgments. These conventions can be difficult to reconcile, making an overall judgment of quality difficult. As a result, the diversity of disciplinary approaches and conventions can be an impediment to innovation in integrative education because such standards may get in the way of developing shared definitions of evidence and quality.

Given the challenges of evaluating interdisciplinary learning, there is scant empirical literature on the career, academic, and personal and interpersonal outcomes of these programs (Borrego et al., 2009; Borrego and Newswander, 2010; Ge et al., 2015; Ghanbari et al., 2014, 2015; Grant and Patterson, 2016), and the existing published research has often used less than rigorous methodological approaches. For example, a prominent feature of the existing research literature on integration is that it often lacks the use of appropriate control groups as a design element. This is not unique to studies of integration; education scholars often struggle in designing studies that examine students in their natural, nonrandom, learning environments (see earlier discussion). Instead, scholars implement the best designs they can to address associations between experiences and outcomes. These designs often, but not always, include the following features: longitudinal approaches, comparison groups, theoretically validated and empirically derived measures of student learning, and samples adequately robust to address the questions asked. Using these design elements as evidence of empirical trustworthiness is a strategy consistent with that offered in the largest and most cited literature synthesis in higher education: *How College Affects Students, Volume 3: 21st Century Evidence That Higher Education Works* (Mayhew et al., 2016). When possible, the committee considered studies that used many or all of these design elements to understand the relationship between integrated experiences and student learning outcomes. However, the generalizability of the student learning outcomes from integrative courses is limited by the fact that no two integrative courses or programs are exactly the same. Each course or program features a different syllabus or curriculum, instructor(s), pedagogical approach(es), class of students, and institutional infrastructure.[5] Additionally, the same course with the same syllabus taught at multiple institutions will have varied results and effects on student learning. There is tremendous variation in how disciplines are integrated, as well as course-to-course variation in teaching and institution-to-institution variation in practice, to name just a few factors. If we use a medical analogy, and consider integration akin to a drug treatment, we see that, at a minimum, the kinds of integration being

[5] This is often true of disciplinary curricula as well.

studied by the committee amount to thousands of different drug treatments. Even defining one of those treatments can be difficult. For example, what is the treatment for a course that mixes the restoration of a motorcycle with literary analysis of *Zen and the Art of Motorcycle Maintenance*? Is the entire course the treatment, is it the mixing of the two fields, is it altered or new teaching practices, or something else? Another challenge is presented by the fact that students may manifest the outcomes of an integrative education experience on a timescale far beyond a semester or other curricular milestone.

THE PATH AHEAD

Despite the many challenges to evaluation and assessment, it *is* possible for faculty and researchers to chart a path toward a meaningful evaluation of an integrative learning experience. The challenge for faculty and scholars who strive to evaluate integrative learning experiences will be to develop frameworks that permit them to evaluate the student learning outcomes they value and hope to provide, rather than those that are easy to measure. This process begins with the faculty member or evaluator asking a series of questions: What are the expected student learning outcomes from this integrative course or program? What is the particular added value to the student experience from this integration? What kinds of methodological approaches will be informative given the learning goals of this integrative education experience? Which methodological approaches will be possible in light of the real-world constraints of evaluating student learning outcomes at my institution? The answers to these questions will necessarily vary according to the instructor, student population served, and specific institutional constraints, even if the same program is implemented across multiple schools in multiple contexts. But the information gained from this introspection can help guide the course or program evaluation appropriately, while also taking into account how access to resources and real-world constraints may shape which research designs are possible and practical. Given the great diversity of integrative approaches in higher education, it is not surprising that a great many diverse learning goals are associated with different courses and programs. Evaluation of these diverse courses and programs will depend on first articulating the expected learning outcomes of the particular course, program, or approach. The practice of "backward design" (described in Box 4-1) could be applied to develop meaningful evaluations of integrative courses and programs. Ultimately, the evidence for integrative approaches in higher education will need to be integrative itself. Causal mechanisms revealed through the use of carefully designed evaluations that feature control groups and longitudinal study designs may be the most informative approach for understanding the

BOX 4-1
Backward Design and Integrative Educational Experiences

"Backward design" is an approach to lesson planning in which curricula are designed by first defining learning goals and objectives, then determining the appropriate assessment tools to evaluate whether those objectives have been achieved, and, only then, designing educational activities to best fit the assessment tools and learning objectives (Graff, 2011; Krajcik et al., 2008; Wiggins and McTighe, 1998). This approach is named backward design because it is the reverse of traditional lesson planning. Because of its emphasis on learning outcomes and evaluation, backward design can serve as a means to figure out the optimal combination of quantitative and qualitative approaches for the evaluation of a given integrative program. First introduced by Wiggins and McTighe in 1998, this approach provides a "roadmap" for instructors who want to focus on the learning goals and objectives, gather the right kind of evidence to demonstrate mastery of those objectives, and create educational experiences to support those learning objectives. Backward design usually involves three stages of planning:

1. **Learning goals and objectives:** what should students understand or be able to do by the end of the experience/course?
2. **Metrics and evaluation:** what kind of evidence or data would demonstrate student mastery of the goals and objectives?
3. **Lesson plans:** what knowledge, experience, or skills will students need to have had in order to perform well on the metrics and evaluations?

In the case of the integration of arts and humanities with STEMM, one could imagine a course in which an instructor had a high-level learning goal for her students to be able to identify specific social impacts of nanotechnology. If she used a backward design approach to build her lesson plans, she might begin by matching curriculum standards at her institution to this broad learning goal in order to create more well-defined learning objectives tailored to the particular needs of her students and her teaching environment. She would then determine the appropriate approach to evaluate her students' mastery of that goal. For example, if one of the specific learning goals involved analyzing how different subcultures (religious groups, socioeconomic brackets, racial groups, etc.) in the United States responded to the same nanotechnology—people's experiences of technology are often shaped by their particular social and cultural backgrounds—she might want students to be able to compare reactions across different subcultures. She might then decide that the best way to evaluate her students' understanding would be through a speaking or writing activity, possibly an oral presentation, which would be graded through a rubric. She would then need to choose lesson activities that best matched the oral presentation evaluative framework. She might schedule time in class for students to do research or write short essays, have guest speakers visit to discuss cultural artifacts from different subcultures, or a classroom activity where students would engage themselves with the piece of nanotechnology in order to gain a better appreciation for its possible impact. This method of designing curricula with the learning goals and objectives as the starting point frees the instructor to pick and choose which combination of metrics and assessments best serve those goals and to design learning experiences aimed at helping students achieve those goals in measurable ways.

impact of one integrative educational experience (and more convincing to one group of stakeholders) (Choy, 2002), whereas the impact of another integrative educational experience might be best understood (and more convincing to another group of stakeholders) through personal narratives of individual success or demonstrations of students' work, as made possible by e-Portfolios (Gulbahar and Tinmaz, 2006). But to move forward with a research agenda will require proponents of integration rooted in the arts, humanities, and STEMM disciplines to come together with each other and with scholars of higher education research to agree on anticipated learning outcomes of integrative courses and programs and to develop appropriate approaches to assessment. The committee anticipates that the publication and dissemination of this report will stimulate this collaborative process, but we also expect that the sustained engagement of these diverse stakeholders over the long term will be necessary.

Multiple goals and student outcomes will be associated with integration, and these goals and outcomes should influence the approach to evaluating integrative programs. To support those efforts and bolster evidence-based evaluative practices, the committee suggests a research program to generate and collect robust forms of the following types of evidence in evaluating the impact of integrative educational experiences:

- Qualitative, longitudinal testimony from students, teachers, and administrators on the impact of integrative programs and courses
- Quantitative, controlled, longitudinal data on the impact of integration on students' grades, attitudes, and competencies, retention and graduation rates, college major, employment status, salary, civic engagement, and satisfaction in life and career
- Narrative case studies that offer an in-depth description and analysis of the nature of integrated programs and courses, the learning goals of the students taking the course, and the goals of the professor teaching the course or program
- Detailed descriptions of the curricula and pedagogies of integrative courses and programs
- Portfolios of student work, performances, or exhibitions

Currently, only limited evidence from each of these categories is available. While this is unfortunate, it is not surprising. Decisions about curricular offerings in higher education are rarely made in response to research studies on the impact of specific educational approaches on students. Rather, decisions about curricular offerings are often driven by the interests and limitations of faculty (e.g., lack of time, greater emphasis on research productivity for professional advancement, etc.) and the cultural and administrative structure of the department or university.[6] Although we urge faculty and administrators to consider research on student learning and development when they make choices about curricular offerings, we kept the reality of curricular decision making in higher education in mind as we crafted the recommendations we put forth in this report. Though we conclude that the evidence on the impact of integrative approaches in higher education is limited, we do not think it is practical for institutions with an interest in adopting integrative approaches to wait for additional research before they begin supporting, implementing, and evaluating integrative approaches, especially if current discipline-based curricular approaches are not serving their students' learning goals.

Given that the limited evidence available is promising and suggests positive learning and career outcomes for students, the committee is urging a new nationwide effort to develop and fund the research needed to establish appropriate protocols for the collection of the kinds of robust and multifaceted evidence that the broader educational community can accept and embrace. Institutions of higher education that have implemented integrative models, or that plan to do so, might consider working with faculty and higher education researchers to evaluate courses and programs in ways that could address a number of outstanding research questions. Among the questions researchers should consider addressing are the following:

- What are the shared or different hypothetical learning outcomes across similar and different integrative courses and programs?
- How might different levels of integration (e.g., multidisciplinary, interdisciplinary, transdisciplinary) impact students differently?
- How might in-course, within curriculum, and co-curricular integrative experiences impact students differently?
- How does integration between closely related disciplines (e.g., the integration of chemistry and biology) impact students compared with integration across more distantly associated disciplines (e.g., engineering and history)?
- What specific mechanisms contribute to the positive outcomes associated with integration?
- What role does pedagogical approach play relative to subject matter?
- How do students make sense of integrative experiences?
- What are the longitudinal and long-term influences of integration?
- Does the integration of the humanities, arts, and STEMM lead to

[6] Lattuca, L. R., and Stark, J. S. (2011). Shaping the college curriculum: Academic plans in context. John Wiley & Sons.

integrative learning? If so, how is integrative learning revealed in student work products?

- Can a successful integrative program or course at one college or university be implemented at another such that it results in the same kinds of learning outcomes? To what extent can the positive impacts of integrative efforts be achieved independently of the idiosyncratic professors and administrators who implement such courses and programs?
- To what extent are students choosing or required to take integrative courses? Courses in their major? Courses outside their major? In other words, what does course taking tell us about the extent to which higher education is siloed?

5

Understanding and Overcoming the Barriers to Integration in Higher Education

As an educational approach, interdisciplinary integration offers many potential benefits (see Chapters 6 and 7), but implementing integrative curricula faces possible barriers at multiple levels, from the preferences and pressures that inform student and faculty choices, to departmental and disciplinary structures, to college- or academy-wide priorities and practices, such as budgeting practices and criteria for promotion and tenure review. These challenges are practical and organizational, but they are also *cultural*: established practices remain so in part because professional identities, disciplinary structures, and organizational and bureaucratic arrangements are interlinked in ways that tend to sustain the status quo (Crow and Dabras, 2014). At the same time, marking these challenges simply as "cultural," where culture stands for the inertia or recalcitrance of established arrangements, risks giving too crude of a diagnosis. It is important, therefore, to undertake a more nuanced examination of the factors that might inhibit integration and to explore how these barriers can be overcome.

In this chapter we offer an in-depth discussion of the common barriers to integration, while also acknowledging that, as the many examples offered in this report demonstrate, some faculty have found ways to develop integrative courses and programs, and some institutions have facilitated the implementation of these curricula. As such, they are exceptions to the status quo that prove that barriers to integration are not insurmountable. Though the committee does not believe there is a one-size-fits-all approach to overcoming barriers to integration, here we offer readers a practical framework for developing implementation strategies that can be catered to the local context of their specific institution. We also describe several examples of how certain institutions have overcome some of the common barriers to integration.

THE INSTITUTIONAL BARRIERS TO INTEGRATION IN HIGHER EDUCATION

The National Academies and others have devoted significant prior attention to both the potential of interdisciplinary research and teaching to advance human knowledge in new and innovative ways as well as the factors that inhibit it (NRC, 2003, 2004). Today, we see that interdisciplinary classes and team teaching are relatively common. A national survey conducted by the Higher Education Research Institute at the University of California, Los Angeles, found that greater than 40 percent of full-time faculty who teach in 4-year institutions reported that they had taught an interdisciplinary class within the past 2 years (Jacobs, 2013, p. 198). This remarkable level of interdisciplinary instruction is not a new phenomenon; it has been a consistent pattern since enthusiasm for interdisciplinary approaches began in the 1990s. Yet, despite enthusiasm for interdisciplinary approaches in teaching and research, it has also been well documented that numerous factors tend to discourage interdisciplinary integration—even within related fields. The vast majority of courses, even if cross-listed, are discipline-specific and not integrative: their content and learning goals are recognizable as particular to a single discipline, and they are taught by faculty members with training in that discipline who hold a teaching position in a disciplinary department (Jacobs and Frickel, 2009). Notwithstanding significant investment in advancing interdisciplinarity within the natural sciences and engineering over the past several decades, and in spite of widespread receptivity of such initiatives, many of the barriers to integration that were enumerated over a decade ago in the National Academies 2005 study *Facilitating Interdisciplinary Research* persist today. While the focus of previous National Academies studies has been on interdisciplinary integration within the science, technology, engineering, mathematics, and medicine (STEMM) fields, the insights that apply to interdisciplinarity, in general, also apply to the integration of the arts, humanities, and STEMM fields.

Many of the barriers to interdisciplinary teaching derive from an array of established arrangements and practices that tend to keep significant departures from internally controlled disciplinary norms in check (Abbott, 2001). In other words, some of the most significant barriers to integration derive from entrenched practices that, although not built to intentionally discourage such alternative approaches, do nevertheless have that effect. Thus evaluating the potential benefits of integration also invites us to reflect on the potential opportunity costs and impediments to change associated with discipline-centric curricular, professional, and organizational approaches in higher education.

Sociologist Andrew Abbott (2001) has observed that a dual institutional structure has contributed to the continuity of disciplinary departments in American universities over the past century. The structure of disciplinary (departmentally administered) majors came into being very soon after the discipline-department structure itself, and has remained equally as durable. He notes: "Once institutionalized, the major system has never been questioned. Indeed, it has never really been the subject of a serious pedagogical debate, since allocating the undergraduate curriculum on some basis other than majors raises unthinkable questions about faculty governance and administration" (Abbott, 2001, p. 128). On the macro level, disciplines shape labor markets for academic faculty. Positions tend to be within disciplines, and, in general, careers tend to remain within the bounds of a discipline rather than the bounds of a single university. This structure is then reflected in the organization of disciplines into university departments. With a few exceptions, most American universities have the same mix of departments, with broader professional disciplinary structures and individual university departments perpetuating each other. As already noted, the overwhelming majority of faculty hold Ph.D.s in the same field (i.e., from the same sort of department) as the department in which they hold teaching positions. Even as new interdisciplinary fields (e.g., communications, cognitive science, bioinformatics) or professional or applied programs (e.g., business) have emerged, they have tended to follow the same structural patterns as the traditional disciplines. For example, initially highly interdisciplinary, a majority of faculty in communications departments now also have Ph.D.s in communications (Jacobs and Frickel, 2009). The mutually sustaining relationship between professionalized discipline and department affects the arrangement of everything, from the allocation of resources to faculty hiring, graduate training, and the characteristics of undergraduate training.

Discipline-based departments do not preclude the development of interdisciplinary activity, but they do regulate it. Disciplinary structures persist as centers of gravity that generally ensure that interdisciplinary efforts tend to exist on the margins of established disciplines and are often less sustainable and hold less institutional agenda-setting power than disciplines.[1] Thus, even though there may be interest in pushing research and teaching

[1] The sociology of education literature notes that the exceptions to the marginalization of integration and interdisciplinarity come about when (1) big money pours in from somewhere outside the university and helps create an interdisciplinary field that persists so long as the money does and (2) social and political developments create demand that universities feel compelled to respond to—e.g., women and gender studies, Afro-American studies, and science, technology, and society.

in interdisciplinary directions, there is a pull toward disciplinary structures that discourages all but the least risk-averse from pursuing such trajectories. Rhoten and Parker (2004) have observed that younger researchers are more interdisciplinarily inclined than senior researchers, yet are also more aware of potential costs associated with interdisciplinary work, which include risks to employability within the academic job market or in achieving tenure based on disciplinarily defined criteria of quality. Interestingly, Rhoten and Parker also reported that researchers in their study who chose to pursue a riskier, interdisciplinary professional course did so out of the belief that such a course was more likely to contribute to important societal needs, even if potentially at a cost to their own professional success (2004). The faculty who spoke with the committee over the course of this study reported similar motivations and constraints. For the most part faculty told the committee that they developed integrative courses in spite of, not because of, departmental and disciplinary priorities. And despite investing significantly more time and energy than would have been required to teach a traditional, nonintegrative course, they generally did not (and did not expect to) receive extra recognition from their departments. This investment of time and energy came at the cost of developing a more conventional portfolio of research and teaching within a single discipline. Indeed, at many institutions faculty may be unable to count integrated course development and implementation toward their career advancement, and if integrated courses require significantly more time and labor than simply reproducing a standard, disciplinary course, faculty may not be able to afford the professional cost of such efforts. Reappointment, promotion, and tenure criteria often require faculty to meet teaching, research, and service expectations, all of which tend to be evaluated according to disciplinarily defined criteria and by senior members of their discipline. Thus, even though early career faculty are more inclined to take their research or teaching in more interdisciplinary directions, they are also the most vulnerable to the potentially negative professional consequences of doing so (National Academy of Sciences, National Academy of Engineering, and Institute of Medicine, 2004).

Even where there may be an explicit institutional support for developing integrative courses, disciplinary control over the evaluation of faculty work nevertheless poses risks that may lead faculty to be conservative. Mansilla and Gardner have shown that researchers evaluating interdisciplinary work tend to rely on proxy criteria (patents, citation counts, journal impact factors, etc.) that reflect tacit disciplinary priorities. Thus, even where interdisciplinary work is supported in principle, it may be penalized in practice because of the application of inappropriate criteria in evaluations of quality (2006). Mansilla and Gardner observed that researchers doing interdisciplinary work "were often critical of these 'proxy' criteria because they saw them as ultimately representing a disciplinary assessment of their interdisciplinary work."

These barriers and negative incentives persist regardless of whether integrative courses prove to be more or less effective than the traditional alternatives. It is simply much less risky and much less demanding for faculty to avoid innovative teaching practices and instead to continue doing what has been done before. Even simply asking whether there is a better way tends to be inhibited. And if the question cannot be asked in a serious and sustained way, neither will it be possible to achieve a robust and definitive answer. Thus, while questions remain about the extent of the benefits of integrative learning, disciplinary and institutional conservatism exerts a strong counterforce to the sorts of experimentation in integration that would help answer these questions.

Another way that disciplinary structures exert this conservative effect is in the sheer effort required for faculty trained in those disciplines to acquire the opportunities and competencies necessary to teach integrative courses. Disciplinary and departmental structures are not only epistemological structures but also social and, in certain respects, physical structures. At many institutions, humanities departments are located near each other but far from natural science and engineering departments. Departmental meetings, committee service, and other opportunities for interaction often privilege interactions with faculty in the same discipline. Unless it is actively encouraged, there may be very limited occasions in which faculty from disparate fields might meet and interact with each other, let alone find themselves in circumstances where they might discover points of convergence between their teaching goals, interests, or methods (Feller, 2002; Rekers and Hansen, 2015).

Even for the faculty who actively seek out opportunities or partnerships for integrative teaching, doing so requires investing time and labor. One of the challenges of offering integrative courses is the lack of faculty preparation and expertise for offering them. In an environment in which faculty tend to be trained by, and squarely located within, a discipline (and corresponding department), efforts at achieving interdisciplinarity in research and education are difficult, and are initially more likely to be aggregative than integrative; that is, they will tend to draw dollops of material from the various different pools of disciplinary expertise in parallel rather than bring them into conversation or productive confrontation. Chapter 3 of this report categorized this form of integration through aggregation as multidisciplinary rather than interdisciplinary. Multidisciplinary courses offer the opportunity to juxtapose complementary skills and content areas, and in some cases teaching a multidisciplinary course might serve as a stepping-stone to full interdisciplinarity, which requires the development of a sense of points of connection and coherence, and also of disjunction—

of differences in underlying epistemological assumptions, methodological commitments, and styles of thought (Crombie, 1994; Fleck, 1981; Hacking, 1992). However, for faculty from distinct disciplines to co-teach an interdisciplinary course requires the time and freedom to work through different assumptions and goals in the search for common ground and synthesis that transcends their disciplinary methods and styles of thought. Though rare, such opportunities can engender in faculty, as well as students, the capacity to achieve a sense of the critical limitations of a given disciplinary approach and recognize and explore the unoccupied intellectual territories and potentially transformative, yet unasked, questions that lie beyond the existing borders of disciplines.

There are, however, some categories of faculty whose training makes them particularly well suited to teach integrative courses without requiring the extra resources needed for two faculty from traditional disciplines to co-teach a single course. For example, some faculty have been trained in interdisciplinary fields that exist at the intersection of disciplines, such as Arts, Media, and Engineering, or Science and Technology Studies (Jasanoff, 2010). They engage in research that is already integrative and thus are well positioned to introduce students to the forms of knowledge and the methods in which they are expert. Yet early-career scholars with such training tend to have a difficult time finding positions, particularly in traditional disciplines, and may feel pressure to shift their research, publishing, and teaching practices to reflect the norms of their discipline or department. Even where there is an openness within traditional departments to incorporate interdisciplinary courses into the curricula of the major, if these courses are to exist, there must be positions and support mechanisms at institutions of higher education for faculty willing and able to teach such courses.

The departmental structure is, of course, only one node in a larger network of structures, all of which shape and are shaped by each other. Academic departments interested in fostering interdisciplinary integration may still face institutional or academy-wide challenges, for instance in justifying interdisciplinary courses in budgetary allocations or in incorporating them into curricula where accreditation standards disallow or appear to discourage departures from traditional, disciplinary practices. Also, the structure of majors and the tendency of advisers and departments to (often incorrectly) characterize the major as a straightforward path to a particular profession, may encourage students to think of coursework outside the major as unnecessary and extraneous to their education, even as frivolous. Given that, for the vast majority of students, the cost of higher education is a significant concern, if they do not have a sense of the potential value of coursework that is not marked as valuable by virtue of being a requirement of the degree, they are unlikely to avail themselves of opportunities to take integrative courses, even if such courses exist. Thus the majors system can inhibit the success of interdisciplinary courses, not only by imposing certain requirements on students but also influencing how students form impressions of what is valuable (and not) in their education in terms of those requirements. For students to recognize the potentially significant value of academic exploration, particularly of the sorts evaluated in this report, there must be a corollary commitment within institutions of higher education to inform students of the value of such exploration.

OVERCOMING THE BARRIERS TO INTEGRATION IN HIGHER EDUCATION

Overcoming the barriers to integration requires more than simply providing endorsement of, or even resources for, integration (Klein, 2009). It also requires that the variety of existing structures and practices that discourage departures from disciplinary norms be discerned, evaluated, and actively mitigated. This report catalogs and describes a diverse array of integrative programs and courses that demonstrate that overcoming some of the common barriers to integration is possible. These existing courses and programs can serve as models and offer lessons learned about how to implement integrative courses and programs; however, the committee believes that no one-size-fits-all "solution" exists for overcoming the barriers to integration. Although the common barriers to integration are widely shared among institutions, solutions for overcoming those barriers will almost certainly apply differently depending on local circumstances. Institutional leadership, modes of resource allocation, existing departmental and divisional policies, and other factors will all have a significant impact on whether and how an integrative course or program is successfully implemented. For instance, at some institutions deans can promise faculty that their tenure requirements will be tailored to their field in specific ways, while at others this would be unthinkable. The same variation applies to how courses get approved, cross-listed, and incorporated into majors curricula, and how teaching is assessed (e.g., by student feedback, by number of courses taught, "service" teaching, new course development).

Nevertheless, the committee recognizes that many faculty and leaders will look to this report for insight into how they can successfully implement new integrative educational experiences on their campuses. To those stakeholders looking for practical input on how to implement integrative courses and programs we offer not a specific list of solutions but rather the description of a general process that can guide the implementation of an integrative course or program. The specific context of a particular institution will determine what form implementation will take in practice. This process unfolds in five major stages:

1. Articulate the goals and intended outcomes of an integrative educational experience.
2. Assess the institutional context—the opportunities, constraints, and assumptions.
3. Identify in your institution's curricular framework the opportunities where integrative learning can enhance students' learning goals.
4. Consider existing best practices at your institution (if any) and some possible practices at other institutions that might potentially achieve the goals and intended outcomes of integration in the context of your institution.
5. Use a design process that includes a mix of idea shaping, testing of strategies, outcomes assessment, and iteration to successfully implement a new integrative course or program.

In many ways, the structure of this report reflects this process. Chapters 1, 2, 6, and 7 each speak, in different ways, to the many goals and intended outcomes of integrative educational experiences, while the foregoing section of this chapter and portions of Chapter 2 assess some of the common constraints and assumptions that influence the institutional context for integration. However, steps 3, 4, and 5 of the process are necessarily highly context dependent. There is enormous variety among institutions of higher education in the United States, and approaches that have proven successful at one institution may not succeed at another. The following sections offer several examples of institutions that have recognized the interconnections between the common barriers to integration described above and have sought to develop strategies that address them. While we cannot say that these strategies will be applicable at all, or even most, institutions, they can serve as examples of "possible practices" for implementing integration. We do not offer these examples as solutions, per se, but as conversation starters that will stir the pot, spark the imagination, and inspire faculty to ask further questions.

EXISTING PRACTICES AIMED AT OVERCOMING COMMON BARRIERS TO INTEGRATION

As we have already described, promotion and tenure criteria can serve as barriers to integration. In an effort to overcome this common barrier, some institutions have begun to change policies to explicitly include interdisciplinary scholarship within the criteria for promotion and tenure. For example, the University of Michigan is explicit that faculty should "receive full credit for their contributions to interdisciplinary and/or collaborative scholarly projects." Similarly, Indiana University Bloomington's tenure guidelines state that "candidates for tenure and promotion are encouraged to pursue innovation wherever it seems promising, even at the edges of disciplinary boundaries or in between them." Rochester Institute of Technology (RIT) has also made changes to its governance documents for faculty retention, promotion, and tenure to explicitly authorize and encourage boundary-crossing scholarship—categories of work that are now explicitly recognized as scholarship for purposes of promotion and tenure include "scholarship of integration" in which "faculty use their professional expertise to connect, integrate, and synthesize knowledge."

In the case of RIT, these changes to the tenure and promotion criteria are part of a larger effort to make interdisciplinarity an explicit element of a university-wide strategic plan to "diminish the lingering effects of a silo culture." The plan specifically calls for encouraging interdisciplinary collaboration and learning, including by reducing impediments for approval of jointly offered interdisciplinary programs and inviting students to develop individualized, interdisciplinary undergraduate projects and courses of study. RIT leadership has recognized that impediments to interdisciplinary integration are cultural, institutional, budgetary, and logistical. Thus it calls for innovation in budgeting procedures, in processes for allocating space, and in local decision making that can facilitate generative risk taking.[2]

The University of Arizona (UA) offers another example of an institution that has taken steps to facilitate integration. For example, UA has developed Graduate Interdisciplinary Programs (GIDPs) that bring faculty together to offer Ph.D. programs in interdisciplinary fields that cross multiple disciplinary boundaries. Sixteen GIDPs offer Ph.D. degrees or minors. Some of these GIDPs cross boundaries of STEM, social sciences, and humanities. For instance, the Second Language Acquisition and Teaching GIDP is associated with five colleges and 19 participating departments, including Anthropology, East Asian Studies, Linguistics, and Speech, Language and Hearing Sciences. In addition, Cognitive Science, an interdisciplinary study of human mental processes, includes pursuit of problems in reasoning, language comprehension, and visual recognition and involves the integration of disciplines such as philosophy, psychology, neuroscience, linguistics, and computer science. In addition, graduates of the Arid Lands Resource Sciences GIDP work in interdisciplinary fields that include international development; famine, famine early warning systems, and food security; land use, history, change, degradation, desertification, management, and policy; ethnoecology and other ethnosciences; economic and agricultural policy and development; borderlands issues; globalization; civil conflict; and urban development as they relate to the arid and semiarid lands of the world

[2] See https://www.rit.edu/president/strategicplan2025/dimension5.html (accessed September 21, 2017).

Part of the success in overcoming some of the common barriers to interdisciplinary teaching has come from giving GIDPs a certain amount of autonomy and budgetary independence, as well as a role in faculty promotion and tenure. GIDPs are administered by the graduate college and an executive committee of program faculty. The graduate college provides administrative support and moderate levels of teaching assistantship support. The institutional budget model rewards departments and colleges by crediting faculty teaching in a GIDP course with the student credit hours and thus awarding the relevant tuition income to the college of the instructional faculty member. Furthermore, the GIDP has a voice and a role to play in the promotion and tenure process of the faculty members participating in the GIDP.

Arizona State University (ASU) offers another interesting example of an institution that has taken steps, at multiple levels—from restructuring of majors to the organization of departments and schools—to promote greater integration across disciplines. One way that ASU has encouraged the development of integrative courses is to make them a required part of the curriculum of certain majors. For example, undergraduates at ASU pursuing bachelor of science degrees are required to complete six credit hours (generally, two courses) in "science and society." The purpose of this requirement is to expose students to the reciprocal relationships between science and society, to develop a critical understanding of the scientific principles and problems underlying the scientific dimensions of issues of significant importance in the public domain, and to cultivate students' capacities to formulate, communicate, and defend well-reasoned views about such issues. As a result of this requirement, thousands of students at ASU take courses that, in one way or another, seek to integrate arts, humanities, and/or social sciences with STEM.

The science and society requirement came into being in conjunction with broader reforms of disciplinary and curricular structures at ASU. In the past 15 years, most departments at ASU have been restructured into transdisciplinary schools that combined faculty from a variety of disciplines to achieve forms of interdisciplinary breadth capable of addressing complex "grand challenges" (Crow and Dabars, 2015). This process began with creating the School of Life Science, which merged several previously separate life science departments, and integrating faculty positions (and, therefore, course offerings) for historical, philosophical, ethical, and social scientific research related to the life sciences. The intraschool integration of biological sciences with life sciences–focused humanistic and social scientific research has allowed forms of integrative teaching, research, and graduate training that are generally difficult to achieve when spanning separate units. Fifteen years later, courses that integrate the biological sciences with humanistic study of biology's historical, social, and ethical dimensions have become a routine element of undergraduate biology education at ASU and have been taken by thousands of students. Since then, other units at ASU have also been restructured in similar ways, facilitating the development of a variety of integrative undergraduate and graduate programs. To offer just a few examples, the degrees offered by the School of Sustainability draw upon fields as diverse as environmental sciences, political science, economics, sociology, and ethics. The School of Arts, Media and Engineering combines faculty in fields as diverse as computing, design, mechanical engineering, philosophy, and media arts, and offers courses and degrees that reflect this diversity. A majority of biology undergraduates take writing-intensive biology and society courses, and doctoral committees for the Ph.D. in Biology & Society routinely include a combination of natural scientists, social scientists, and humanists.

Although ASU offers an example of particularly radical and rapid restructuring of traditional disciplinary and departmental arrangements, and its approach is, therefore, unlikely to be readily replicated at other institutions, it does offer some important lessons. Opportunities for integrative education have followed from commitment to fostering integrative interdisciplinary research and by creating faculty units that are themselves integrated by virtue of combining arts, humanities, and STEM expertise. Teaching, training, and research that are collaborative between faculty with backgrounds in humanities and natural science fields happen much more readily because numerous barriers are eliminated. Whereas at many institutions it can be difficult for faculty from different departments to get permission to team-teach a course, it is no more difficult for an ecologist and an environmental ethicist who are in the same school at ASU to co-teach a course than for an evolutionary biologist and a geneticist. Collaborative teaching and research tend to happen more spontaneously because faculty from different disciplines may interact more frequently and informally simply by virtue of occupying the same building; serving together on administrative, hiring, and doctoral committees; and encountering each others' research in the context of departmental business like promotion and tenure reviews. Each of these settings has tended to offer opportunities for faculty to encounter and understand the approaches of colleagues from other disciplines, laying valuable but otherwise difficult to achieve foundations for offering integrative learning opportunities to students. Because integration is part of the culture of the unit, integrative courses are integrated into the curricula of majors, and an interdisciplinary faculty is present to advise students, direct honors theses, and so forth. Students who are not otherwise convinced of the need to look for learning opportunities beyond a disciplinary major will nevertheless encounter them.

The strategies at Arizona State University, University of Arizona, Rochester Institute of Technology, University of Michigan, and Indiana University Bloomington that we describe here offer a few examples of how specific institutions have worked to break down the cultural and administrative barriers to integration. These examples were chosen not because they are superior to other efforts under way at other institutions, but simply because they are models with which the committee is familiar. Indeed, several of these examples are from institutions with which members of the committee are affiliated. We hope that these examples offer readers of this report a sense for how institutions *can* facilitate integration; however, we do not expect that the strategies that have worked at these schools will necessarily work at other institutions. We urge those interested in facilitating greater integration at their institution to consider these examples, along with additional existing strategies in place at other institutions, as a source of inspiration that might inform the development of approaches to integration that are catered to the specific goals, needs, and constraints of their specific institution.

6

The Effects of Integration on Students at the Undergraduate Level[1]

The case for integrating the arts, humanities, and science, technology, engineering, mathematics, and medicine (STEMM) fields in higher education ultimately must rest on evidence. The committee was charged with examining the evidence behind the assertion that integrative experiences in the arts and humanities and STEMM lead to educational and career successes. The two important questions raised by this charge are: what constitutes *success*, and *how* can this success be measured?

These two questions lead to a host of more detailed questions. What are the learning objectives of integrative courses and programs? Is it to learn content knowledge in more than one discipline? To learn content knowledge in one discipline better or differently? To learn how to learn in an entirely new way? To learn to see connections between disciplines and integrate two or more areas of knowledge? To better prepare a graduate for employment in a particular sector or to enhance lifelong habits of learning? What kinds of outcomes constitute success? If it is impossible to assess the outcomes of a program or course using a controlled experimental design, is it still worth evaluating and will the emerging evidence be informative?

These questions are not easy to answer and may be answered differently by different disciplines, faculty members, institutions, and scholarly journals. As such, evaluations of programs and courses, if they are done at all, may be very distinct and difficult, or impossible, to generalize (see

[1] The Committee wants to acknowledge and thank research consultant Matthew Mayhew for his significant contributions to this chapter. A commissioned literature review written by Dr. Mayhew on behalf of the committee contributed directly to the writing of this chapter.

Chapter 4 for a discussion of "The Challenges and Limitations of Research on Integration").

The research literature we review in this chapter is limited in several ways. Very few studies used designs that account for selection effects through randomized, controlled trials or quasi-experimental designs; employed longitudinally administered, theoretically valid, and empirically tested measures of student learning; and applied data control, collection, and analytic methods consistent with the effort's theoretical underpinnings. Moreover, given that each course and program is unique and that relatively few have been studied, it is difficult to generalize the student outcomes associated with a particular integrative educational experience. Further, the studies reviewed here did not fully take advantage of qualitative and narrative approaches to assessment that might offer insight into those student outcomes that are difficult or impossible to capture quantitatively (see Chapter 4 for an extended discussion).

That said, it is important to note that the challenge of designing rigorous evaluations within the context of real-world courses and programs is not unique to studies of integration. Most higher education research on student learning struggles with challenges related to the generalizability of student learning outcomes associated with a particular educational approach and with evaluation design in a real-world context. Instead, we considered multiple forms of descriptive, qualitative, and quantitative evidence.

In this chapter we describe the committee's approach and the broad conclusions we drew regarding undergraduate student learning outcomes associated with integration of the arts and humanities into STEMM curricula, and vice versa. We offer an overview of some existing studies that have examined student outcomes associated with certain models of integration. When qualitative and quantitative research studies were available, we focused our analysis on the outcomes of these studies. However, when this kind of research was not available, we sought to analyze programs and courses based on other sources of information (e.g., descriptions of courses, student work, scholarly output, etc.). When possible, we point out promising models and practices for the design, implementation, and evaluation of integrative courses and programs. We highlight these models such that institutions that are interested in developing new integrative courses and programs can draw inspiration from existing efforts and can modify and transform these models to suit the specific learning goals of their students.

THE COMMITTEE'S APPROACH

Despite the challenges to assessment of integrative courses and programs described in Chapter 4, the committee worked to "examine the evidence behind the assertion" that integration leads to improved educational and career outcomes by collecting and considering evidence from multiple sources.

To collect the evidence for the report, the committee engaged in the following activities:

- Held several meetings to learn from the literature and from each other;
- Heard from the public through open sessions of committee meetings;
- Learned from noted experts in relevant fields who met with the committee;
- Conducted careful literature reviews (both commissioned externally and through the Academies' library professionals and staff);
- Examined examples of integrative programs from across the country, including examples garnered from a "Dear Colleague" letter calling for input from faculty; and
- Evaluated other information relevant to the effort.

In the spirit of integration, the committee examined diverse forms of evidence for the study, including:

- Personal testimony from faculty, administrators, students, and employers on the value of an integrative approach to education;
- Essays and thought pieces that make logical arguments for integration based on observations and evidence on the state of higher education today;
- Descriptions of integrative courses and programs;
- Formal and informal evaluations of courses and programs carried out by institutions; and
- Peer-reviewed literature from the field of higher education research.

Although there are limitations to the evidence base on the impact of integrative programs and courses on students, we found that the available research does permit several broad conclusions to be made.

- Aggregate evidence indicates that certain approaches that integrate the humanities and arts with STEM have been associated with positive learning outcomes. Among the outcomes reported are increased critical thinking abilities, higher order thinking and deeper learning, content mastery, creative problem solving, teamwork and communication skills, improved visuospatial reasoning, and general engagement and enjoyment of learning (see Tables 6-1 and 6-2 for an overview of the learning outcomes associated with specific integrative approaches).

- The integration of STEM content and pedagogies into the curricula of students pursuing the humanities and arts may improve science and technology literacy and can provide new tools and perspectives for artistic and humanistic scholarship and practice.
- Many faculty have come to recognize the benefits of integrating arts and humanities activities with STEM fields and can testify to the positive learning outcomes associated with integrative curricula.
- Abundant interest and enthusiasm exists for integration within higher education, as evidenced by the groundswell of programs at colleges and universities in various sectors of American higher education (see "Compendium of Programs and Courses That Integrate the Humanities, Arts, and STEMM" at https://www.nap.edu/catalog/24988 under the Resources tab).

DOES THE DIRECTION OF THE INTEGRATION MATTER?

The statement of task for this study asked the committee to consider courses and programs that integrate "STEMM curricula and labs into the academic programs of students majoring in the humanities and arts" and integration of "curricula and experiences in the arts and humanities into college and university STEMM education programs." In some instances, the direction of the integration—whether STEMM into the arts and humanities or arts and humanities into STEMM—is clear, but in many other instances the integration is more of a balanced, mutual integration. This led the committee to question whether distinguishing the direction of the integration matters for this analysis. The committee concluded that it may matter only in cases where an integrative course is available or advertised primarily to students pursuing a specific major (for example, an engineering and design course that is offered only to engineering students), or when the goals of a course and intended learning outcomes are primarily rooted in one discipline (for example, an engineering and design course for which knowledge and skill in engineering are the primary goal of the course).

In the sections below, we describe the results of the committee's research into integrative courses and programs. The courses and programs we highlight below are not intended as a comprehensive list, but rather as examples of existing courses and programs—some of which have been in existence for some time (for a more comprehensive list of known integrative programs and courses see "Compendium of Programs and Courses That Integrate the Humanities, Arts, and STEMM"). When possible, we highlighted programs and courses for which there is published research or a formal or informal evaluation.

We separate this discussion into two main sections that correspond to the direction of the integration; however, it could be argued in many instances that a particular course or program we describe is a mutual, balanced integration rather than integration with a clear direction. We point out several such examples in each section but maintained this organizational structure because we observed that integrative courses that were explicitly offered to all students, as opposed to students from a particular discipline, were rare.

INTEGRATION OF THE ARTS AND HUMANITIES INTO THE ACADEMIC PROGRAMS OF UNDERGRADUATE STUDENTS MAJORING IN STEM

Several initiatives have been used to increase the quality of college-level STEM education over the past 30 years, including the integration of arts and humanities courses with STEM curricula—a movement that has gained more attention in recent years (Borrego and Newswander, 2010; Catterall, 2012; Dail, 2013; Ge et al., 2015; Grant and Patterson, 2016; Maeda, 2013; Sousa and Pilecki, 2013). Based on the premise that adding concepts and pedagogies from the arts and humanities to STEM courses will increase creativity (Jones, 2009), deepen understanding of content and increase knowledge retention (Asbury and Rich, 2008; Jeffers, 2009; Land, 2013), support critical thinking and problem solving (Lampert, 2006; Spector, 2015), and make learning more fun and engaging (Brown and Tepper, 2012), many advocates for "STEAM" education have pushed for innovative curricular and co-curricular integrative and interdisciplinary educational opportunities for students at all levels of education (Borrego and Newswander, 2010; Grant and Patterson, 2016; Maeda, 2013; Sousa and Pilecki, 2013; Spector, 2015).

Below we offer an overview of the existing research on student outcomes associated with the integration of the arts and humanities into STEM courses and curricula. Though many of the studies we review suffer from methodological limitations, taken together they do suggest that integrative courses (in-course integration) and integrative programs (within-curriculum integration and co-curricular integration) are associated with positive student outcomes, including higher order thinking, creative problem solving, content mastery of complex concepts, enhanced communication and teamwork skills, and increased motivation and enjoyment of learning (Gurnon et al., 2013; Ifenthaler et al., 2015; Jarvinen and Jarvinen, 2012; Pollack and Korol, 2013; Stolk and Martello, 2015; Thigpen et al., 2004). Notably, many of these learning outcomes are consistent with the "twenty first–century professional skills" that employers say they are looking for in recent graduates (see Chapter 2).

In-Course Integration

The committee's review of the literature turned up many examples of in-course integration (see "Compendium of Programs and Courses That Integrate the Humanities, Arts, and STEMM"—available at https://www.nap.edu/catalog/24988 under the Resources tab, a subset of which have associated student learning outcomes available in the higher education research literature. Such courses can be extremely diverse and may integrate the arts, humanities, and STEMM fields in varying degrees. As the following examples demonstrate, in-course integration can range from a course that includes a small component of another discipline (e.g., a neuroscience course with an assignment involving haikus) to a fully integrated course (e.g., a design engineering course). When embedded in coursework, integrative mechanisms and processes are associated with positive learning outcomes (see Table 6-1).

For instance, a comparative study of an undergraduate neuroscience course by (Jarvinen and Jarvinen, 2012) found that students who were required to apply their understanding of neurotransmission through the creative activity of making a 3- to 5-minute film significantly outperformed those who learned the concept from more conventional approaches. The authors also found that this learning transcended several levels of Bloom's revised taxonomy. In addition, students who participated in the integrative assignments reported that, while it was challenging to simplify the process of neurotransmission into a video, they felt more confident in their ability to apply neurotransmission in future classes. The process of creating helped them reduce the complexity of the scientific concept to its most salient features.

Conveying scientific content with accuracy requires deep understanding of the concepts being conveyed. This depth of knowledge comes from internalizing information and constructing it into a form that is unique and coherent to the individual. Pollack and colleagues developed assignments that use the writing of haiku—a 17-syllable poem—as a means for students to identify key neurobiological concepts and to articulate them in a succinct yet creative manner (Pollack and Korol, 2013). Using student questionnaires, Pollack and colleagues found that the haiku writing process and explanations created a context for deconstructing complex concepts into simple terms and reconstructing those concepts to produce descriptions that reflected deep meaning. The haiku assignments fostered logical thinking skills and guided students to understand that claims need to be supported by evidence that is, in turn, synthesized by the student's reasoning (Pollack and Korol, 2013).

In both of the courses described above, students demonstrated higher order thinking, as defined by Bloom's revised taxonomy, and were better

TABLE 6-1 Learning Outcomes from In-Course Integrative Programs

Title (Author)	Integration Level	Specific Androgogical Features	Key Measured Learning Outcomes and Higher Order Skills Addressed
Elevating Student Potential: Creating Digital Video to Teach Neurotransmission (Jarvinen and Jarvinen, 2012)	Course-Level Assignment	Undergraduate students create a 3- to 5-minute film to display their understanding of neurotransmission.	• Improved content mastery • Improved extension and application • Analysis and synthesis
The Use of Haiku to Convey Complex Concepts in Neuroscience (Pollack and Korol, 2013)	Course-Level Assignment	Undergraduate students create a haiku-style writing to articulate key neurobiological concepts such as addiction.	• Deconstruction of complex concepts • Evidence support of claims • Creative thinking • Synthesis and deep meaning
Integrating Art and Science in Undergraduate Education (Gurnon et al., 2013)	Course-Level Assignment	Undergraduate students create a 3-dimensional sculpture based on protein-folding research.	• Improved intuition for complex concepts of protein structure and folding • Interpretation of scientific data • Creative thinking
Can Disciplinary Integration Promote Students' Lifelong Learning Attitudes and Skills in Project-Based Engineering Courses? (Stolk and Martello, 2015)	Full Course Integration: Direct comparison of outcomes in an traditional materials science course and an integrated course in materials science and history	Project-based learning in both courses. Students completed the Situational Motivation Scale (Guay et al., 2000) and the Motivated Strategies for Learning Questionnaire (Pintrich et al., 1991) and also self-report on critical thinking, self-efficacy, and the value of the learning tasks.	Students in the integrated course show: • Increased motivation and engagement in self-regulated learning • More frequent use of critical thinking skills • Higher self-efficacy • Higher value placed on learning tasks

continued

TABLE 6-1 Continued

Title (Author)	Integration Level	Specific Androgogical Features	Key Measured Learning Outcomes and Higher Order Skills Addressed
A Model for Teaching Multidisciplinary Capstone Design in Mechanical Engineering (Thigpen et al., 2004)	Full Course Integration: Interdisciplinary course from the departments of electrical engineering, marketing, and art.	Capstone course designed to transfer academic skills to the workplace. Focus on teamwork, transitioning from classroom to industry and product design, manufacture, and marketing.	Not an empirical study. Authors report: • Improved communication skills • Improved insight into practical aspects of workplace engineering • Improved multidisciplinary teamwork • Improved employment opportunities
Exploring Learning: How to Learn in a Team-Based Engineering Education (Ifenthaler et al., 2015)	Course-Level Assignment in Designing for Open Innovation Course	Individual and team assignments, class discussion.	• Increased positive attitude toward the engineering course • Increased confidence for performing in the course • Improved team dynamics
Arguments for Integrating Arts: Artistic Engagement in an Undergraduate Foundations of Geometry Course (Ernest and Nemirovsky, 2015)	Full Course Integration	Activity-based foundations of geometry course for secondary mathematics educators using geometry software, physical devices, artistic engagement activities, field trips, and written reflection.	• Synthesis—an improved ability to blend mathematics with other life experiences • Creative thinking—developed innovative ways of fostering mathematical inquiry • Increased appreciation of art

able to communicate effectively the complex ideas associated with neuroscience. Students creating the videos also showed increased engagement in the material, regardless of personal career interests (Jarvinen and Jarvinen, 2012; Pollack and Korol, 2013).

Similar learning outcomes were observed by faculty in a DePauw University course that integrated visual arts with biochemistry through sculpture-building based on protein-folding research. According to Gurnon and colleagues, students were able to develop "an intuition for complex concepts of protein structure and folding" (Gurnon et al., 2013, p.3).

Courses that integrate the arts, humanities, and STEMM fields are also associated with increased student motivation and engagement. Faculty at the Olin College of Engineering offered two options to students taking an introductory materials science course: an integrated materials science-history course co-taught by faculty in engineering and history, or a nonintegrated course taught only by an engineering professor (Stolk and Martello, 2015). Although both courses were project based and had similar structures, students who participated in the integrated course demonstrated increased motivation and engagement in self-regulated learning strategies over the term compared with students in the nonintegrated course, as measured by the Situational Motivation Scale (Guay et al., 2000) and the Motivated Strategies for Learning Questionnaire (Pintrich et al., 1991). Additionally, students in the integrated course self-reported using critical thinking skills in their work more frequently and had higher self-efficacy and valuing of learning tasks than students in the nonintegrated course.

In addition to using creative assignments in established disciplinary courses, some faculty have designed new courses based on integration. For example, mechanical engineering students at Howard University have the option of enrolling in a multidisciplinary capstone course with students from the departments of electrical engineering, marketing (in the business school), and art (in the Division of Fine Arts). Although the faculty who coordinate this course do not provide empirical evidence supporting the relationship between student participation in this initiative and learning and career outcomes, they argue that students "gain insight into the practical aspects of engineering in the workplace, develop skills in working on multidisciplinary teams, experience a transitional step between classroom and industry, gain an understanding of how the curriculum is relevant to real world product design, manufacturing and marketing, develop and improve communication skills and, most importantly, improve their opportunities for employment" (Thigpen et al., 2004, p. S2G-1-6 Vol. 3).

The use of integrated approaches with prospective instructors can have a multiplier effect on future teaching. For example, prospective secondary mathematics teachers enrolled in an activity-based foundations of geometry course were taught the "synthetic and analytic aspects of projective geometry through the use of physical devices and dynamic geometry software" (Ernest and Nemirovsky, 2016, p. 5) as well as artistic engagement activities, such as "exploring the roots of projective geometry in Renaissance art, participating in whole-class discussions and composing written reflections on ideas in the arts, visiting a contemporary art museum, and creating individual artistic pieces using ideas from projective geometry" (p. 6). Qualitative results indicate that the students in the course demonstrated the ability to blend mathematics with other life experiences, could identify innovative ways of fostering mathematical inquiry, and shifted their attitudes toward art.

Summary of Student Outcomes Associated with In-Course Integration

In-course integrative initiatives, whether by including an arts and humanities–related assignment in a disciplinary STEM course or through implementation of a fully integrated course, have an association with student learning outcomes. Although there are several limitations to the generalizability of the research presented above, the evidence suggests that in-course integration shares a relationship with higher order thinking, creative problem solving, content mastery of complex concepts, enhanced communication and teamwork skills, and an increased engagement of learning. Further research is needed to fully understand the relationship between these student outcomes and integrative educational experiences.

Within-Curriculum Integration

The committee also considered a number of examples of within-curriculum integration. As the examples below will demonstrate, participation in interdisciplinary curricula has been associated with similar positive learning outcomes to those observed in in-course integration. In particular, as shown in Table 6-2, within-curriculum integration is associated with critical think-

TABLE 6-2 Learning Outcomes from Within-Curriculum Integrative Programs

Title (Author)	Integrated Disciplines and Curricula	Specific Androgogical Features	Key Measured Learning Outcomes and Higher Order Skills Addressed
Integrated Curricula: Purpose and Design (Everett et al., 2000) The Freshman Integrated Curriculum at Texas A&M University (Malavé and Watson, 2000)	First- and second-year foundational engineering curriculum components from calculus, chemistry, engineering graphics, English, and physics—horizontally integrated. Upper-level curricular elements vertically integrated	Ethics, writing, graphics, problem solving. Students held accountable for all disciplinary components in all courses. Force Concept Inventory, Mechanics Baseline Test, and California Critical Thinking Skills Test	• Improved critical thinking skills • Improved calculus and physics performance • Higher GPA • Improved computer skills • Improved facility for teamwork • Higher retention rates for underrepresented engineering students

Title (Author)	Integrated Disciplines and Curricula	Specific Androgogical Features	Key Measured Learning Outcomes and Higher Order Skills Addressed
The Effect of a First-Year Integrated Engineering Curriculum on Graduation Rates and Student Satisfaction: A Longitudinal Study (Olds and Miller, 2004)	*Connections* program designed to highlight the importance of the first-year engineering curriculum by developing significant links between science, humanities, and engineering	Integrated program modules and active learning strategies, interdisciplinary seminars and peer study groups	• 25% improvement in graduation rates • Self-reported post-graduation improvements in critical thinking skills, clarity of contextual significance, ethical awareness, and communication skills • Increased first-year student retention rates • Improved overall student satisfaction
Motion Picture Science: A Fully Integrated Fine Arts/STEM Degree Program (Scholl et al., 2014)	Fully integrated BFA in Film and Animation with BS in Imaging Science	Components of each degree program completely integrated to form one, new undergraduate degree	• 96% of graduates obtained positions in their fields • Graduates attributed career success to diverse skill sets
Learning Across Disciplines: A Collective Case Study of Two University Programs (Ghanbari, 2015)			Program directors report • Improved knowledge retention • Improved learning enjoyment • Broadened perspectives No control group and no quantitative measures

ing skills, content mastery, facility to work in teams, and communication skills (Malavé and Watson, 2000; Olds and Miller, 2004; Willson et al., 1995). Within-curriculum integration is also associated with higher GPA and improved retention and graduation rates (Everett et al., 2000; Malavé and Watson, 2000; Olds and Miller, 2004).

An early program integrating STEMM and arts and humanities was founded in 1994 at the Dwight Look College of Engineering at Texas A&M University (Everett et al., 2000; Malavé and Watson, 2000). This program, which came out of the National Science Foundation's 1993 Foundation Coalition for Engineering Education, integrates the first-year components of calculus, chemistry, engineering graphics, English, physics, and problem solving into a "cross-discipline engineering, science, and English curriculum" (Everett et al., 2000, p. 172). Faculty also developed integrated second- and upper-year models focused on presenting a unified approach to the engineering sciences and discipline-specific specialization, respectively. As Everett et al. explain, "One can view the first and second year models as performing horizontal integration, building a wide, highly interconnected foundation onto which the upper division builds vertically" (Everett et al., 2000, p.168). The first-year curriculum included areas such as ethics, writing, graphics, problem solving, physics, calculus, and chemistry. As a highly coordinated integrated curriculum, students were held accountable in all courses for information presented in any one of the other disciplines, which means they needed to know the material from their English course to do well in their physics course.

Coordinators of this program used control groups and longitudinal design to assess the impact of this program on student outcomes. Using instruments such as the Force Concept Inventory (Hestenes et al., 1992), the Mechanics Baseline Test (Hestenes and Wells, 1992), and the California Critical Thinking Skills Test (Facione and Facione, 1992), researchers found that students who participated in the Foundation Coalition first-year integrated program demonstrated better critical thinking skills, performed better in calculus and physics, exhibited higher overall GPAs, developed significantly better computer skills, and expressed greater facility to work in teams than students who completed the traditional first-year curriculum (Malavé and Watson, 2000; Willson et al., 1995). Of particular note, Foundation Coalition students who identified as underrepresented in engineering had higher retention rates than similar students in the traditional curriculum.

A program similar to Texas A&M's Foundation Coalition initiative was designed at the Colorado School of Mines in 1994. The *Connections* program, which also was developed in response to calls for engineering education reform in the mid-1990s, was designed to help students form connections in their first-year courses and "understand the importance of their first-year studies by allowing them to develop appropriate and significant links among disciplines" (Olds and Miller, 2004, p. 25). Students who participated in the program enrolled in science and engineering courses where faculty used integrated project modules and active-learning strategies, participated in a two-semester interdisciplinary seminar that "further developed and explored the interconnectedness of appropriate topics from each of the first-year science, humanities, and engineering courses" (p. 25), and engaged in peer study group systems. Olds and Miller, 2004 found that "average" engineering students who participated in this program graduated at rates approximately 25 percent higher than students in the traditional curriculum (Olds and Miller, 2004). Additionally, through a follow-up survey 5 years later, these students indicated that their experience in *Con-*

nections enhanced their academic preparation by helping them make connections among course topics, improving their critical thinking abilities, setting a context for their science and engineering studies, increasing their awareness of ethical issues, and strengthening their communication skills. In addition, Olds and Miller noted that resources spent to have top faculty reach and mentor first-year students resulted in increased retention and overall satisfaction with the educational experience.

Within-curriculum efforts also have sought to educate students at both ends of the "STEM-Art spectrum" (Scholl et al., 2014, p. 2). The Rochester Institute of Technology initiated an undergraduate program in Motion Picture Science, for example, that fully integrates components of a bachelor's in fine arts in Film and Animation and a bachelor's of science in Imaging Science into one undergraduate degree. Scholl and colleagues (2014) reported that 96 percent of students who completed this program received positions in motion picture or imaging fields upon graduation, and many of these graduates attributed their career success to the diverse set of skills and techniques learned in the program (Scholl et al., 2014). Additionally, in her qualitative, collective-case study of two arts and STEM integrated programs at two universities—the ArtScience and ArtTechnology programs (Ghanbari, 2014, 2015)—Ghanbari attributed knowledge retention of course concepts, a rise in enjoyment of learning, a broadening of perspectives, and a substantial influence on future careers to the experiential collaborative learning dimension of the integrated experience.

Less is known about the impact of fully integrated arts, humanities, and STEMM majors. Although students in the Motion Picture Sciences program at the Rochester Institute of Technology demonstrated high career placement rates, the authors did not compare these students to others with similar ambitions in other programs. Ghanbari's study of the ArtScience and ArtTechnology programs also did not include a control group, nor did it quantitatively measure the impact of student participation. As with in-course integration, more rigorous research is needed to fully assess the influence of within-curriculum integrative initiatives on student learning and career outcomes.

Summary of Student Outcomes Associated with Within-Curriculum Integration

Within-curriculum integrative initiatives at colleges and universities share positive associations with student learning outcomes. For example, when compared with non-participant peers, students who participated in the first-year integrated engineering programs at Texas A&M University (Everett et al., 2000; Malavé and Watson, 2000; Willson et al., 1995) and Colorado School of Mines (Olds and Miller, 2004), which integrated

STEM disciplines with English courses and humanities concepts (e.g., ethics), had higher retention and graduation rates, stronger critical thinking skills, increased subject matter competence in their science and engineering courses, and improved communication skills. Many of these student outcomes are shared by the other within-curriculum programs described above (see Table 6-2). Since the literature on the Texas A&M program provides extensive detail as to how the Freshman Coalition curriculum was developed and assessed, it may be a useful resource for others hoping to implement similar initiatives.

Co-Curricular and Extracurricular Integration

Since many co-curricular and extracurricular programs are coordinated outside of the classroom, less is known about their impact on student learning outcomes than is known about curricular interventions. Nevertheless, co-curricular and extracurricular integration between the arts and humanities and STEMM fields has been shown to have positive relationships with student learning (LaMore et al., 2013).

With the rise of makerspaces, collaborative laboratories, and residential learning programs on college and university campuses, as well as student-led clubs and events, students and faculty now have more opportunities to engage in interdisciplinary discovery, enhance disciplinary knowledge and professional goals, and build communities of practice around making. Lewis, for instance, argues that infusing the arts within the STEMM fields through multimedia design studios and makerspaces "has enormous potential to infuse the liberal arts with design thinking, collaboration, creative computing, and innovation while maintaining the level of deep reflection and critical thinking associated with humanist inquiry" (Lewis, 2015, p. 269). However, more research is needed to understand the learning outcomes associated with participation in these co-curricular initiatives within higher education.

One faculty-coordinated co-curricular initiative is the collaborative laboratory setting at Youngstown State University (Wallace et al., 2010). This initiative, which brings together faculty and students from the College of Science, Technology, Engineering and Mathematics and the College of Fine and Performing Arts in a shared and neutral workspace, provides an opportunity for integrated design teams to meet, collaborate, and work on innovative projects. To date, the laboratory has yielded several successful projects, including the development of a retro-styled cell phone prototype designed by mechanical engineering technology and sculpture students. This opportunity to collaborate is thought to have increased the sculpture student's awareness of manufacturing technology, while the engineering students were "challenged to apply engineering principals to address a poorly constrained, creative problem" (Wallace et al., 2010, p. 3E-6).

Another co-curricular integrated program with emerging evidence of success was developed by officials at two museums at Southern Utah University: the Braithwaite Fine Arts Gallery and the Garth and Jerri Frehner Museum of Natural History (Grant and Patterson, 2016). Their partnership, which resulted in several different art and science integrated learning programs for K-12 students, provides undergraduate and graduate art education students with the opportunity to docent during programs and develop lesson plans for museum activities for participating children. So far, Grant and Patterson (2016) have observed learning outcomes related to art education, such as educational program design and management skills.

Summary of Student Outcomes Associated with Within-Curriculum Integration

Though the evidence base is extremely limited, integrative co- and extracurricular activities are common and practitioners report positive outcomes on students. Such programs often take the form of makerspaces, collaborative laboratories, and residential learning programs on college and university campuses, as well as student-led clubs and events. Practitioners argue, based on case studies and their own observations, that integrative co- and extracurricular activities can promote such outcomes as design thinking, collaboration, creativity, innovation, and critical thinking.

EXPOSURE TO THE ARTS CAN SUPPORT SUCCESS IN STEM THROUGH THE DEVELOPMENT OF VISIO-SPATIAL SKILLS

There is evidence to suggest that the integration of certain arts curricula, such as drawing, painting, and sculpting, into the curricula of students can also improve their success in STEM courses by improving their visio-spatial abilities. Uttal and Cohen (2012) present a number of well-controlled, often randomized studies, demonstrating that visuo-spatial ability is highly associated with success in STEM subjects (Mohler, 2007; see also Alias et al., 2002; Deno, 1995; Tillotson, 1984). Some of these studies show that no matter how visual imaging is taught, it has substantive benefits for STEM learning outcomes. Such course material may involve specific visual thinking exercises; consist of learning computer-aided design, or focus on drawing, industrial drawing (or draughting), painting or sculpting, though drawing stimulates ideational fluency over use of computer-aided design programs (Ainsworth et al., 2011; Groenendijk, et al., 2013; Uttal and Cohen, 2012). Indeed, Uttal and Cohen review dozens of controlled studies performed on students ranging from middle-school through graduate school that demonstrate that visio-spatial training intervention, devoid of STEM content, nonetheless results in improved scores on a variety of generalized visio-spatial skill tests and, at the same time, on specific measures of STEM learning such as classroom tests, standardized STEM tests, persistence in major, and probability of graduating within a STEM major. Additionally, groups of students who typically underperform in STEM subjects, such as women and some minorities, benefit more than other groups of students from visio-spatial training (Sorby and Baartmans, 1996, 2000; Sorby, 2009a, 2009b).

INTEGRATION OF STEM INTO THE ACADEMIC PROGRAMS OF UNDERGRADUATE STUDENTS MAJORING IN THE ARTS AND HUMANITIES

The evidence base for the impact on students of courses and programs that integrate STEM knowledge and pedagogy into the arts and humanities is extremely limited, particularly in the peer-reviewed literature. This is unfortunate, as the committee heard passionate testimony from faculty, students, and scholars to the benefits of such integration, both to students and to the arts and humanities disciplines. Corroborating qualitative and quantitative research findings would greatly enrich understanding of the impact of such programs on students and could serve to support the observations and opinions of proponents of this form of integration.

Despite the limited evidence base, we found some evidence that integration of STEMM content, pedagogies, and scholarly approaches into the humanities and arts serves to:

- Improve scientific and technological literacy among students majoring in the arts and humanities
- Offer new tools and approaches for humanistic and artistic scholarship and practice
- Drive artistic and humanistic questioning, scholarship, and practice that explores the influence of science and technology on the human condition

Integration of Stem into the Curricula of Students in the Arts and Humanities Can Promote Scientific and Technological Literacy

The published literature that addresses efforts to integrate STEM into the curricula of students majoring in the arts and humanities has dealt more with the importance of scientific and technological literacy than the value that STEM knowledge and approaches contribute to scholarship in the arts and humanities. Such science and technology literacy courses often endeavor to make STEM content more accessible, relatable, and engaging for students by grounding this knowledge in real-world contexts and demonstrating the impact of STEM on society throughout history and in our everyday lives. As such, these courses integrate information and pedagogical approaches from the arts and humanities with content knowledge and pedagogical approaches in science and technology. Though we treat these courses as examples of STEM integration into the arts and humanities because of the students they are primarily serving, it is also possible to view them as a balanced, mutual integration.

Successful science and technology literacy courses are very popular with students. For example, the University of Virginia course "How Things Work" (Physics 105 and 106) taught by Louis Bloomfield attracted 500 students each semester for more than a decade and has had a significant impact. Enrollment is now capped at 200 students (Krupczak and Ollis, 2006). Many humanities and arts students have not only gained an understanding of physics through this course but also have found the knowledge exciting and useful and are now less intimidated by the discipline. The popularity and impact of this course led to the development of a "How Things Work" textbook that has been used by 200 universities (Krupczak and Ollis, 2006).

Similarly, the course "Science and Technology of Everyday Life" taught by John Krupczak at Hope College has been taken by 1,000 non-engineering students—60 percent of whom are women—since its introduction to the curriculum in 1995. To better understand the impact of this very popular course, Krupczak and colleagues evaluated students' experiences and outcomes using the Motivated Strategies for Learning Questionnaire (MSLQ). The instructors found statistically significant increases in intrinsic motivation, task value, and self-efficacy between the pre-test and post-test across three semesters, as well as a reduction in test taking anxiety (Krupczak et al., 2005). The authors concluded that "the case study shows that non-engineering students can have increased motivation for learning science and technology, increased perceived value for science and technology, increased self-confidence about learning science and technology" (Krupczak et al., 2005, p. SJ1-36).

These are only two examples of many such science and technology literacy courses. Krupczak and Ollis (2006) describe 18 scientific and technological literacy courses at different institutions that are common in their popularity with students and their impact on student's understanding of, and attitudes toward, science and technology and its relevance and impact in society.

Others have noted the value of integrating STEM into meaningful reading and writing assignments as a means of improving science literacy. An article by Glynn and Muth (1994) posits that to achieve science literacy, scientific curriculum should include reading and writing assignments. The authors note that without scientific literacy, students will be underprepared to make informed decisions about scientific or social issues that they confront in their everyday life. Glynn and Muth define meaningful learning as "the process of actively constructing conceptual relations between new knowledge and existing knowledge." They explain that by reading scientific text and endeavoring to write it, students actively familiarize themselves with different concepts and form the foundation of real scientific expertise. As many students, as well as a disproportionate number of women and minorities, are not scientifically literate, reading and writing can serve as an engaging vehicle to learn science meaningfully.

Efforts to integrate engineering knowledge and understanding into the curricula of non-engineering majors have also been a priority of the Sloan Foundation, the Teagle Foundation, and the American Society for Electrical Engineers (ASEE). In 1980, the Sloan Foundation launched the New Liberal Arts Initiative, an effort to encourage a range of diverse institutions to integrate technological and quantitative literacy into liberal arts disciplines (Tobias, 2016). In 2017, the Teagle Foundation, in partnership with the ASEE, launched the Engineering-Enhanced Liberal Education Project that supported an analysis of 16 case studies of courses and programs that colleges and universities had developed in response to the Sloan Foundation initiative (American Society for Engineering Education). The analysis was led by noted author and higher education researcher Sheila Tobias. In her qualitative analysis of the 16 courses, Tobias found "interesting, varied, and so far successful (in terms of faculty commitment and student enrollment) courses and programs" that "encompass a wide variety of initiatives: from the dream and long-term commitment of a single faculty member (Princeton), to the initiative by a college president (Wesleyan), a college-wide commitment to including Engineering in a newly constituted set of General Education offerings (University of Maryland), and to a university-wide one-course Tech graduation requirement at Stony Brook originating in the College of Engineering and Applied Sciences."[2]

Also in the vein of promoting scientific thinking as a competency that all college graduates should possess, the Association of American Colleges and Universities launched the Scientific Thinking and Integrative Reasoning Skills (STIRS) initiative "to develop tools to improve the capacity of undergraduate students to use evidence to solve problems and make decisions." STIRS scholars have developed course modules that "facilitate

[2] See American Society for Engineering Education: https://www.asee.org/engineering-enhanced-liberal-education-project/introduction.

integrative, evidence-based inquiry into real-world problems,"[3] and some of these course modules have integrated aspects of STEM into the arts and humanities. An interesting example from STIRS scholar and Professor of Mathematics at the University of North Dakota Ryan Zerr is a course titled Congressional Apportionment: Constitutional Questions, Data, and the First Presidential Veto. In this course, Zerr uses mathematics and historical evidence to examine how congressional apportionment methods can affect the results of the U.S. presidential election. His course requires students to "describe a variety of mathematical issues that arise in determining how to allocate U.S. House representatives among all U.S. States and use mathematics to develop a notion of fairness regarding apportionment" (Zerr, 2014, p. 1). Zerr suggests that the case study can be used in a first-year seminar course, liberal arts–themed mathematics course, history, political science, or any situation where integrative learning is an objective.

Another STIRS case study is provided by Tami Carmichael from the University of North Dakota Humanities and Integrated Studies Program. The case study uses the topic of the environmental impacts of tar sands oil extraction and transmission to develop student skills in scientific reasoning and critical thinking. Carmichael developed this case study for use in a general education course for students who may not have strong scientific knowledge. Carmichael suggests that the assignments in this case study will require students to "think carefully about the material and use specific data and arguments to formulate reasoned responses."[4]

It is also important to recognize the significant contribution of The Andrew W. Mellon Foundation in supporting interdisciplinary and integrative education and scholarship. As a foundation focused on the humanities and arts, Mellon has invested heavily in efforts to develop curricular interventions at the undergraduate and graduate level, develop centers for interdisciplinary research and education, and support the research and hiring of faculty doing interdisciplinary work. The Mellon Foundation has also heavily invested in the development of interdisciplinary areas of study, including digital humanities; the intersection of architecture, urbanism, and the humanities; environmental humanities; arts and the environment; and medical humanities. The support from the Mellon Foundation has initiated the development of multiple centers for interdisciplinary study and thus is developing the intellectual foundations for further curricular reforms.[5]

---

[3] See Association of American Colleges and Universities Scientific Thinking and Integrative Reasoning Skills (STIRS) at https://www.aacu.org/stirs.

[4] See People, Places, and Pipelines: Debating Tar Sands and Shale Oil Transmission—STIRS Student Case Study at https://list.aacu.org/stirs/casestudies/carmichael.

[5] See The Andrew W. Mellon Foundation at https://mellon.org.

Federal agencies have also supported scientific understanding and engagement through the arts, and in some instances publicly available reports to these agencies by grantees offer some insight into the impact of such efforts. For example, with support from the National Science Foundation (NSF), dancer, choreography, and McArthur Genius Fellow Liz Lerman developed "The Matter of Origins," a multi-media, contemporary performance that explores the beginnings of the universe through art, science, and engagement. In Act One, audience members experience a vivid soundscape and contemporary intergenerational dance based on historical and contemporary understandings of physics. In Act Two, they adjourn to a nearby room to enjoy tea, cake, and dialogue punctuated by dance interruptions designed to stimulate further exploration of the nature of science, spiritual, and scientific explanations of origins, and the limits of scientific measurement.[6] "The Matter of Origins" was performed nine times at three performance sites before audiences ranging in size from 282 to 1,100. Using a mixed methods research design, Diane Doberneck and colleagues evaluated the impact of the performance on audience members and found that "[a]udience members' attitudes, interest, knowledge, and behavior concerning science showed positive change. Quantitative and qualitative data across all three study sites consistently demonstrated these results regardless of audience member demographics or background" (Miller et al., 2011). Although this evaluation was not published in a scholarly, peer-reviewed journal, and though the performance was not a part of a college or university course, it suggests that integration of STEM content and ideas into artistic performances can have positive impacts on public engagement and understanding of science.

The National Science Foundation has offered additional support for integrative and transdisplinary efforts through several funding programs. In 2006, NSF initiated the CreativeIT program to explore "the synergies between creativity and information technology, science, engineering, and design research."[7] The NSF CreativeIT program has funded partnerships between composers and artificial intelligence researchers at Rensselaer Polytechnic Institute to create "a digital conductor of improvised avant-garde performances" (PhysOrg, 2010), research projects focused on the development and application of computational tools and creative methods to problem solving using creative intuition and inquiry, and the use of the performance art of improvisation to develop the creative capacity of individuals and groups in science education and research in the emerging field of computational biology, among many other projects. Research funded by the

[6] See The Matters of Origins Evaluation Study at http://ncsue.msu.edu/research/matterof origins.aspx.

[7] See National Science Foundation at https://www.nsf.gov/cise/funding/creativeit.jsp.

NSF CreativeIT program sought to integrate methods and practices inherent in the fine, applied, and performing arts, computer science, and STEM learning to garner insights and discovery of new knowledge that equally valued creative cognition and computational thinking. NSF's Established Program to Stimulate Competitive Research program has also supported several collaborations between artists and scientists aimed at better understanding and communicating the impacts of climate change. For example, a collaboration between Rhode Island School of Design's Charlie Cannon and Eli Kintisch led to the development of LookingGlass, an augmented reality interface that allows users to view the impacts of climate change on the local environment as if it were happening right in front of them.

Clearly, support from federal agencies and private foundations has had a significant impact on the establishment of new integrative courses, programs, and scholarship. While this national support is necessary for establishing new, innovative integrative teaching and scholarship, it is also important to acknowledge that support from institutions of higher education is also necessary to sustain such efforts.

Stem Integration Can Offer New Tools and Approaches for Humanistic and Artistic Scholarship and Practice

Although the committee could not find published, peer-reviewed research on the impact on students of educational approaches and activities aimed at enriching arts and humanities scholarship by integrating STEM knowledge and pedagogical approaches, descriptions of programs and courses offer some insight into the goals of such integration and its impact on the disciplines. Take, for example, the Digital Humanities, which apply computational methods and data processing to humanistic inquiry (Dalbello, 2011). By applying technology to humanistic research, students are able to approach humanistic scholarship in new ways. For instance, a course at Virginia Tech called Introduction to Data in Social Context offers students the opportunity to examine "the way data is used to interpret patterns of human behavior, identity, ethics, diversity, and interactions."[8] Similarly, Carnegie Mellon University offers an art studio course in computer science titled Electronic Media Studio: Interactivity and Computation for Creative Practice in which students "develop an understanding of the contexts, tools, and idioms of software programming in the arts."[9] The course is open to students who have mastered the basics of programming and would like to use code to make art, design, architecture, and games.

---

[8] See http://ethomasewing.org/idisc_f17/.

[9] See https://ideate.cmu.edu/undergraduate-programs/sound-design/sound-design-course-descriptions.html.

Another example was offered to the committee by Fritz Breithaupt, who elaborated on his *Chronicle of Higher Education* article "Designing a Lab in the Humanities" (Breithaupt, 2017), during a regional information-gathering workshop. In his remarks to the committee and in his article, Breithaupt described how, by working collaboratively with a student with a STEM background, he was able to break new ground in his research into morality narratives and see the similarities between humanistic scholarship and the activities of a scientific research lab. In the article, Breithaupt writes: "The goal of our Experimental Humanities Lab is not to imitate the sciences, but to reclaim what the humanities have always done: Ask questions, observe, question our world, and, yes, experiment and gather data. If that is what happens in a lab, then surely we might have a lab. Why should labs be reserved for the sciences?"

Among the most practical and directly applied intersections of STEM enhancing arts education included courses being taught in Visual Art and Chemistry, Dance and Anatomy and Physiology, Music and Math, and Music and Technology (see "Compendium of Programs and Courses That Integrate the Humanities, Arts, and STEMM" available at https://www.nap.edu/catalog/24988 under the Resources tab). As Robert Root Bernstein, professor of physiology at Michigan State University and arts integrative researcher explains, "All of kinetic art is embedded in engineering training and practices; all of electronic art is embedded in computer technology."

The San Francisco Art Institute (SFAI) requires all students (primarily who are art students) to take science courses. SFAI requires all students to learn quantitative scientific methodologies to develop a scientific mode of inquiry of the world. Many classes at SFAI introduce students to the intersection of art and science. They have numerous arts-sciences courses including: Systems of Investigation: Evolution (Undergraduate Level, Critical Studies), Systems of Investigation: Animal/Human (Undergraduate Level, Critical Studies IN-190-2), Topics in Art & Science: From Miracles to Molecules (Undergraduate Level, Interdisciplinary SCIE-113-1), Life Studies: Biology and Art, Science Deep Time, Vast Space (Graduate and Undergraduate Level, Interdisciplinary), and Studio and Critical Studies Environmental Art and Philosophy.

In its research, the committee also considered courses, programs, and fields of study in the humanities and arts that explore the influence of STEM on society, humanity, and nature. Again, we were unable to learn much about the direct influence of these courses, programs, and fields on student learning outcomes, but in describing them we offer insight into what motivates these activities.

University of California at Davis' 16-year-old experiential Art/Science Fusion Program's mission is to bring the creative energies of the arts and the sciences into a mixture that catalyzes change and innovation in learning for people of all ages.[10]

Cofounders Diane Ullman, entomologist, and Donna Billick, visual artist, developed massive artworks, as well as a number of course such as ENT 001 Art, Science and the World of Insects, SAS 40 Photography: Bridging Art and Science, SAS 42 Earth, Water, Science and Song, Freshmen Seminar 2: Plants in Art and Science, Freshmen Seminar 4 The Face of Darwin, Freshmen Seminar 7: Water in Science and Song, Science and Society 098: Connecting Art & Science: Bringing Environmental Concepts, FRS 002 Bees, Art and Survival and FRS 002: Portraits of the Oak—Exploring the Art/Science Borderland.

A course titled Countertextual Ecologies: Ecopoetics taught by Leonard Schwartz at Evergreen State College offers one example. This course explores "creative and critical approaches to language, with a view to reframing our understanding of the relationship between nature and history" and asks questions like "Is the poem mimetic of nature, or a function of it? How could such a seemingly noble enterprise as 'environmentalism' or 'protecting nature' be problematic? How have powerful environmental imaginaries and narratives served to dangerously simplify how environmental problems and their solutions are conceptualized?"(Evergreen State College, n.d.). Although it is not possible to know the impact of this specific course on students, Evergreen State College, which offers students the opportunity to "connect critical themes across academic subjects," "study subjects in a real-world context," "explore a central idea or theme, team-taught by faculty from different disciplines," and "learn how to be an active, engaged citizen no matter what career you choose" (Evergreen State College, n.d.), also boasts the second fastest time to degree among schools in Washington State (an average of 3.88 years) and the third highest graduation rate among Washington public schools. In addition, it reports that 88 percent and 92 percent of graduates are employed or pursuing graduate or professional degrees within 1 year and 3 years of graduation, respectively. Given that 60 percent of students at Evergreen work while in school, these numbers are striking (Evergreen State College, n.d.). Future evaluation of the impact of the integrative approach taken by Evergreen on student retention, graduation rate, career outcomes, and other measures of student success could offer valuable insights into the potential value of an integrative approach.

Many additional examples are found in the curriculum at Worcester Polytechnic Institute (WPI), another school that has embraced an integrative approach. In addition to courses that integrate the humanities and arts into engineering, WPI offers courses that integrate STEM concepts

---

[10] See http://artsciencefusion.ucdavis.edu (Accessed November 1, 2017).

and practices into humanistic and artistic contexts. For example, in the course Making Music with Machines and Musical Robotics, taught by Scott Barton, students explore "aesthetic and technical considerations of physical automatic mechanical (electro)acoustic instruments and the music that they make" and design and build new machines to make new kinds of music.[11] Another course, The Philosophy of Technology, taught by John Sanbonmatsu, "considers the epistemology, phenomenology, ethics, and politics of technology" and asks students to consider questions like "Is technology value neutral? Or does it have a politics? What makes one technology 'appropriate,' another technology anti-democratic or dangerous? Have we lost control over our technologies? Do computers have a gender? Is technology an artifact, a social practice, or a way of being-in-the-world? All three? Is virtual reality changing what it means to be human? How should our technological artifacts be developed? Should some not be developed at all?"[12]

Experiential learning, engaged learning, and community-based learning strategies, while not new to higher education, are gaining traction on college and university campuses. The Ohio State University STEAM Factory Idea Foundry located in downtown Columbus is 60,000 square feet of workshops and offices, working nooks, classrooms, and communal spaces.[13] It is both a physical place and the belief that each of us has the potential to bring our ideas to life if given the space, the equipment, and the support to empower our inner maker. On and off our campuses we have seen the rapid rise of transdisciplinary centers and institutes, maker spaces and collabs for creativity, innovation, and discovery. (See "Compendium of Programs and Courses That Integrate the Humanities, Arts, and STEMM" available at https://www.nap.edu/catalog/24988 under the Resources tab—which includes a short list of centers and institutes identified as "new" or in the making in the past few years).

Other institutions that have embraced integrative approaches to higher education and offer courses and programs focused on the integration of STEM into the arts and humanities, and vice versa, include Arizona State University (ASU) and the Massachusetts Institute of Technology (MIT).

ASU's President, Michael Crow, describes the university in the following way: "We do things differently, and we constantly try new approaches. Our student's paths to discovery don't have to stay within the boundaries of a single discipline. Our researchers team up with colleagues from disparate

---

[11] See https://web.wpi.edu/academics/catalogs/ugrad/mucourses.html (Accessed December 1, 2017).

[12] See https://web.wpi.edu/academics/catalogs/ugrad/pycourses.html (Accessed December 1, 2017).

[13] See https://steamfactory.osu.edu/ (Accessed December 1, 2017).

fields of expertise. We use technology to enhance the classroom and reach around the world."[14] ASU has created faculty positions for humanities scholars within science and engineering departments such as the School of Life Sciences and the School of Biological and Health Systems Engineering that offer integrative courses to a large number of science and engineering majors. For example, the Biology and Society faculty in the school of life sciences teach courses to biology majors that integrate the life sciences with various humanistic approaches. ASU created a "Science and Society" requirement for bachelor of science students and supports faculty positions for teaching integrative courses that fulfill this requirement. In practice, that means that thousands of ASU STEM undergraduates take integrative, writing-intensive courses at the intersection of arts, humanities, and STEM disciplines.

The MIT Media lab, founded in 1980, "continues to check traditional disciplines at the door. Product designers, nanotechnologists, data-visualization experts, industry researchers, and pioneers of computer interfaces work side by side to invent—and reinvent—how humans experience, and can be aided by, technology."[15] Current MIT Media Lab projects include research on the power of virtual reality to enable new methods for storytelling, engagement, and empathy through a virtual reality narrative film called "TreeSense" (Liu and Qian, 2017) and the development and prototyping of conducive, temporary tattoos called "DuoSkin" that allows users to control their mobile devices, display information, and store information on their skin while serving as a statement of personal style (Kao, 2017). While studies on the impact of the MIT Media lab on student learning are not available, for four decades the Lab has graduated students who have gone on to successful careers at the intersection of art and technology. MIT reports that Media Lab alumni have started 150 spinoff companies (MIT Media Lab, n.d.).

Though ASU, MIT, WPI, and Evergreen all offer courses that integrate STEM into the arts and humanities, each of these institutions also offers courses that integrate the arts and humanities into STEM, though it is challenging to assign a direct impact on student outcomes to these courses (see "Compendium of Programs and Courses That Integrate the Humanities, Arts, and STEMM" available at https://www.nap.edu/catalog/24988 under the Resources tab).

---

[14] See https://asunow.asu.edu/20160912-asu-news-asu-selected-nations-most-innovative-school-second-straight-year (Accessed December 1, 2017).

[15] See https://www.media.mit.edu/about/mission-history/ (Accessed December 1, 2017).

THE IMPORTANCE OF MOVING FROM ANECDOTE TO EVALUATION

Anecdotes, course descriptions, and the testimony of faculty offer meaningful information about the nature of integrative efforts, the goals for student learning embedded in integrative curricular approaches, and the observed or hypothetical outcomes associated with integrative courses and programs, and they should not be dismissed out of hand. As we discuss in Chapter 4, evidence is always collected in stages and all discoveries begin with observations. However, in order to have confidence in the impact of an integrative course or program on a student, anecdotes and other forms of descriptive evidence should be used as the basis for designing methodologically rigorous qualitative and quantitative evaluations.

The importance of evaluating student outcomes is emphasized when we consider studies that have not supported an expected benefit of integration. For example, many are familiar with the "Mozart effect," in which listening to music is believed to improve student performance on mathematics tests (Jashke, 2013; Mehr et al., 2013); however, several studies have not supported this hypothesis. For instance, Mehr et al. (2013) found that "overall, children provided with music classes performed no better than those with visual arts or no classes on any assessment." These results speak to the importance of not assuming that any integrative effort will necessarily yield the expected student outcomes and the importance of viewing anecdote and observation as a starting point, rather than an end point, for understanding impact.

KEY INSIGHTS FROM A "DEAR COLLEAGUE" LETTER

In an effort to collect additional evidence and input, the committee shared a letter with the membership of the Association of Public and Land-Grant Universities, the American Association of Community Colleges, the American Association of State Colleges and Universities, the American Council on Education, the Association of American Universities, the National Association of Independent Colleges and Universities, the AAC&U, and the Alliance for the Arts in Research Universities. We received 79 responses from faculty members, administrators, and non-academic respondents from this query. In addition, members of the committee further solicited their membership and network of colleagues for additional information, to identify additional courses, programs, and initiatives not represented in the peer-reviewed literature. Though this was not a formal scientific survey, nor was it intended to be, the feedback the committee received offered a useful snapshot of the kinds of programs and practices in place at various institutions, the motivations for implementing integrative programs at various institutions, and the challenges faced by institutions that have implemented integrative courses and programs. Respondents shared with the committee a range of qualitative, quantitative, and anecdotal information. The committee is grateful to all who responded to the letter. (A capture of this information titled "Compendium of Programs and Courses That Integrate the Humanities, Arts, and STEMM" is available at https://www.nap.edu/catalog/24988 under the Resources tab).

A review of the 79 responses revealed that respondents cited positive impacts of integrative experiences on students, including student appreciation, real-world relevance, and the opportunity to think critically across disciplines. Some respondents also attributed interdisciplinary exposure to helping students make more informed choices about their undergraduate major or minor. It is possible that the positive accounts we received on the impact of integrative courses and programs represent a positive response bias among respondents; and so, as stated above, it is important that readers not take the information summarized here as the results of a survey or other formal research endeavor. However, the committee chose to describe in this report the input it received from the "Dear Colleague" letter because the observations made by respondents on the nature of the impact of integrative programs on students, as well as the descriptions of formal and informal evaluations of programs, offer useful insight into how some faculty and institutions are defining, approaching, and evaluating integrative efforts, the rationale for integration on various campuses, and the challenges and barriers to integration. As we explain in Chapter 4, all discoveries begin with observation. The observations made by the stakeholders who responded to this "Dear Colleague" letter could form the basis of hypotheses that could be researched in a formal way in the future.

Most responses to the letter indicated that faculty or departments have conducted some form of informal and/or formal assessments to measure impacts, including student performance, and to receive student feedback. Informal assessments often take the form of course evaluations (using surveys or questionnaires). No respondents cited negative assessment results. Several respondents indicated that faculty and departments have gone beyond student questionnaires and have worked with professional assessment centers to measure specific learning outcomes. The responses indicated that while most faculty will take it upon themselves to conduct assessments, some departments and institutions intentionally include assessment plans as part of curriculum development.

Three programs provided examples of assessments done by integrative programs. The first example comes from Portland State University (PSU). At PSU, the 4-year general education program, the University Studies Program, is required of all students with the exception of those enrolled in Liberal Studies or the Honors Program. University Studies begins with Freshman

Inquiry, a year-long course introducing students to different modes of inquiry and providing them with the tools to succeed in advanced studies and their majors. At the sophomore level, students choose three different Sophomore Inquiry courses, each of which leads into a thematically linked, interdisciplinary cluster of courses at the upper-division level. Finally, all students are required to complete a capstone course that consists of teams of students from different majors working together to complete a project addressing a real problem in the Portland metropolitan community. Mentoring sessions, workshops, and seminars are also built into the program.[16]

University Studies' assessment practices have been recognized nationally. In 2010, the Council for Higher Education Accreditation awarded PSU the CHEA Award for Outstanding Institutional Practice in Student Learning Outcomes, based in large part on the assessment practices in University Studies.[17] In 2007 the Association of General and Liberal Studies recognized PSU with the Award for Improvement of General Education.[18]

All Freshman Inquiry courses are co-taught by three to four faculty members. Freshman Inquiry faculty wanting to propose a new course offering must explicitly demonstrate the interdisciplinary aspect of the course.[19] For example, in the Design and Society course proposal, faculty indicated the following: "Our faculty team has expertise in architecture and architectural history, art history, landscape design, structural engineering, studio art, theater history, electrical engineering, and semiconductor physics. Other disciplines represented in the course materials (see Section IV) include social history, film, economics, business ethics, ecology, product design, and industrial design." This course was proposed in 2004 and is still being taught.

PSU conducted a formal assessment of the Freshman Inquiry for the 2014-2015 school year.[20] This assessment employed two methods: an end-of-year survey (809 students responded), which asked students to rate their experiences in their Freshman Inquiry course, and an e-portfolio Review (257 student portfolios were randomly selected), which scored student portfolios against rubrics developed to measure student learning related to the goals of the University Studies Program. The results of the assessment indicated that, in general, students agreed that they had opportunities to address all four of the University Studies' goals in their Freshman Inquiry courses, including inquiry and critical thinking, communication, diversity of human experiences, and ethics and social responsibility. The e-portfolio

[16] See https://www.pdx.edu/unst/home (Accessed December 1, 2017).

[17] See http://www.learningoutcomeassessment.org/Award-WinningCampuses.htm (Accessed December 1, 2017).

[18] See http://www.agls.org/exemplaryprogram.htm (Accessed December 1, 2017).

[19] See https://www.pdx.edu/unst/freshman-inquiry (Accessed December 1, 2017).

[20] See https://www.pdx.edu/unst/university-studies-goals (Accessed December 1, 2017).

analysis revealed that 79 percent of Freshman Inquiry students met program expectations for writing performance.

Lyman Briggs College (LBC) was founded in 1967 at Michigan State University (MSU) to bridge the gap between the sciences and the humanities described by British scientists and novelist C. P. Snow in his 1959 Rede Lecture (Snow, 1959).[21] About 625 first-year students self-select into the program, and all freshmen are required to live in the residence hall where LBC classrooms, laboratories, faculty, staff, and administrative offices are located. The college uses teaching techniques that help students develop writing, reasoning, and presentation skills; focuses on the research culture of science; is dedicated to meaningful student-faculty and student-student interactions; and has smaller introductory science courses to foster a more inclusive environment. Introductory laboratories employ inquiry-based experiments, and a wide range of co-curricular activities expand the ideas learned in class to new and personal settings.

Three aspects of LBC particularly shape its interdisciplinary approach: the absence of departments, strategic space allocation, and faculty joint appointments. Faculty members are drawn from across the natural and social sciences and are hired in mixed groups, which encourages early cooperation and allows members to learn more about a colleague's field. Faculty act as a single governing body, and although they form disciplinary groups based on subject, these groups are explicitly required to work together, and proposals that involve multiple disciplines receive priority. Diverse academic offices are placed side-by-side, allowing for greater interaction between faculty from different backgrounds. Almost every tenure-system faculty member at LBC has a joint appointment in a disciplinary-based department elsewhere on campus, thereby promoting communication and exchange with the rest of the colleges.

LBC has a first-year retention rate of 95 percent, a 6-year graduation rate of 85 percent, and a STEM retention rate of 70 percent, which is substantially higher than the 50 percent national average. In a 2012 survey completed by 466 LBC students, 96.8 percent and 73.3 percent indicated that class size and inquiry-based laboratories, respectively, added either "a great deal" or "a moderate amount" to their LBC experience (Sweeder et al., 2012). The vast majority—92.8 percent—indicated that their STEM courses had "a great deal" or "a moderate amount" of influence on their performance in upper-level STEM courses in their major. For both the class size and preparation questions, female students were significantly more likely to indicate a greater positive response. Evaluations also have shown that LBC students consistently earn higher grades in organic chemistry, bio-

[21] See https://lymanbriggs.msu.edu/news_and_events/2017/LBC2Cultures.cfm (Accessed December 1, 2017).

chemistry, physiology, microbiology, and genetics. Of the 115 senior-level respondents, 48.7 percent had conducted research with a professor outside of a laboratory course, 11.3 percent had coauthored a publication with a faculty member, 38.3 percent had participated in a study abroad program, and 24.3 percent had worked as an undergraduate learning assistant.

In the College of General Studies of Boston University, the sophomore year curriculum includes a two-semester Natural Science sequence that is taught by teams of three faculty—a natural science professor, a humanities professor, and a social science professor—for two semesters.[22] Because the faculty share the same students for a year, they make connections among their courses. For instance, the faculty member who teaches environmental ethics in the sophomore ethical Philosophy Class (HU 201 and 202) schedules her lectures and discussions to coincide with the environmental science units being discussed in Natural Science 202, so that the courses reinforce one another and students can reflect on the interrelationship of the subjects. The sophomore year culminates with a capstone project in which students work in groups of five to six on a 50-page, group-written research project that explores and poses a solution to a contemporary real world problem.

Boston University conducted a 2-year assessment of this program, funded by the Davis Educational Foundation, to evaluate the general education work posted by 100+ students on their e-portfolios between 2011 and 2013. In assessing this work, a team of 11 faculty used a rubric based on AAC&U models to gauge students' levels of competency in areas such as critical thinking and perspective taking, writing skills, and awareness of historical and rhetorical contexts. Student performance was assessed in seven key learning outcomes over four terms in the program. The seven outcomes were: written and oral communication; analyzing and documenting information; awareness of specific historical literary and cultural contexts; rhetorical and aesthetic conventions; critical thinking and perspective taking; integrative and applied learning; and quantitative methods. The results of the assessments showed that students on average make 22 percent to 33 percent progress in these learning outcomes areas over their four semesters in the program, which is significantly higher than the 7 percent progress that national studies of freshmen and sophomores have shown.[23]

While the assessment resulted in encouraging quantitative and qualitative data from reviews of 106 e-portfolios, the respondents reported that the department continues to face challenges in terms of faculty and student buy-in. Some students are not posting work from each of their classes, and some faculty are not encouraging them to do so. The department also faces

---

[22] See https://www.bu.edu/academics/cgs/ (Accessed December 1, 2017).
[23] See http://bu.mcnrc.org/bu-oa-story/ (Accessed December 1, 2017).

challenges in translating assessment data into curricular and pedagogical change.

Barriers to Implementing Integrative Programs and Courses

In the "Dear Colleague" letter, the committee asked respondents to provide input on the obstacles to integration at their institution. Specifically, the committee asked respondents: "Are there factors at your institution that make integration across disciplines difficult to achieve? If so, what are they?" Though all of the respondents reported positive impacts of an integrative approach at their institution (many of which we describe in this report), they also reported systemic cultural and administrative barriers to implementation. Among the common barriers reported were:

- Institutional leaders and faculty members lack a commitment to integration.
- Institutional leaders and faculty members lack time to implement integrative approaches.
- Faculty are dis-incentivized to collaborate due to budgetary issues.
- Faculty are isolated in their own disciplines.
- Faculty are reluctant to collaborate in interdisciplinary work (for example, in engineering and business departments).
- General education programs are self-limiting, not allowing for new, innovative courses to be added to the program.
- Traditional divides across the arts and sciences result in funding inequities.
- "Siloing" of an integrative course within a single department leads it to be overlooked by other relevant disciplines.
- Administrative guidance and funding are inadequate.
- Faculty members in the sciences would like their students to take fewer humanities courses.
- It is difficult to introduce additional STEM content because each discipline writes its own requirements for the degree.

These responses, which revealed similar barriers to integration across a variety of different contexts and institutions, offer valuable insight into how doing integration can surface challenges and barriers that may not to be evident when staying within disciplinary boundaries. Further, the rich input the committee received in response to the "Dear Colleague" letter highlights how collecting evidence about integrative activities can help to illuminate not only the potential benefits of integration for student learning outcomes, but also the sorts of institutional factors that encourage or inhibit integrative approaches.

THE IMPACT OF INTEGRATION ON GROUPS UNDERREPRESENTED IN THE SCIENCES AND ENGINEERING

Improving the representation of women and underrepresented minorities in STEM is a national priority. In our analysis of the evidence on the impact of integrative educational programs, the committee found several instances in which the integration of the arts and humanities with STEM was associated with particular benefits for women and underrepresented minorities. For example, in the Stolk and Martello study that evaluated the impact of an integrative engineering and history course, the authors found that women in the course reported more "significant motivational and self-regulated learning gains" compared with the men in the course (Stolk and Martello, 2015, p. 434). Also, University of Rhode Island's successful International Engineering Program (IEP), in which engineering students double major in a foreign language and an engineering discipline (coupled with a study abroad experience), reported that "women have enrolled in engineering in increasing numbers" (Fischer, 2012). Furthermore, the Foundation Coalition first-year integrated program, which integrates the first-year components of calculus, chemistry, engineering graphics, English, physics, and problem solving into a "cross-discipline engineering, science and English curriculum," reported that students who identified as underrepresented in engineering had higher retention rates than similar students in the traditional curriculum (Everett et al., 2000; Malavé and Watson, 2000; Willson et al., 1995). Also, when Union College's Computer Science Department changed its introductory-level curriculum to be more integrative, focusing more on real-world issues and highlighting humanistic themes such as games and creativity (Union College, n.d.; Settle et al., 2013), it saw an increase in the number of women enrolled.

Further, research has demonstrated that the integration of certain arts curricula, such as drawing, painting, and sculpting, can have particular benefits for women and underrepresented minorities in STEM. Rigorous research has shown that exposure to certain arts curricula can improve visio-spatial ability, which is highly associated with success in STEM subjects (Ainsworth et al., 2011; Groenendijk et al., 2013; Uttal and Cohen, 2012). Indeed, dozens of controlled studies performed on students ranging from middle-school through graduate school have demonstrated that visio-spatial training interventions result in improved scores on a variety of generalized visio-spatial skill tests and, at the same time, on specific measures of STEM learning such as classroom tests, standardized STEM tests, persistence in major, and probability of graduating within a STEM major. These studies find that women and some minorities benefit more than other groups of students from visio-spatial training (Sorby, 2009a, 2009b; Sorby and Baartmans, 1996, 2000).

Taken together, these outcomes are significant because they suggest that an integrative approach to STEM education could be one avenue toward improving the representation of women and minority groups in STEM.

SUMMARY OF THE EVIDENCE FROM UNDERGRADUATE PROGRAMS AND COURSES

The evidence reviewed in this section demonstrates that integrative experiences in college can enhance student learning and development. Despite the lack of strong causal evidence to support the assertion that integration leads to improved educational and career outcomes, the aggregate evidence reviewed by the committee shows that certain educational experiences that integrate the arts and humanities with STEM at the undergraduate level are associated with increased critical thinking abilities, higher order thinking and deeper learning, content mastery, creative problem solving, teamwork and communication skills (Gurnon et al., 2013; Ifenthaler et al., 2015; Jarvinen and Jarvinen, 2012; Malavé and Watson, 2000; Olds and Miller, 2004; Pollack and Korol, 2013; Stolk and Martello, 2015; Thigpen et al., 2004; Willson et al., 1995). Additional outcomes specifically associated with programs that integrate the arts and humanities with engineering at the undergraduate level include higher GPAs, retention rates, and graduation rates (Everett et al., 2000; Malavé and Watson, 2000; Olds and Miller, 2004).

Most of the evidence relating to student outcomes came from studies of programs that integrated the arts and humanities into engineering courses (and to a lesser extent science courses). Much less evidence is available on courses and programs that integrate STEM content and pedagogies into the curricula of students majoring in the humanities and arts. As this chapter demonstrates, this is not due to a shortage of courses and programs that integrate STEM into the humanities and arts. Indeed, entire disciplines, such as Science, Technology, and Society, Bioethics, and Human–Computer Interaction, as well as whole university departments, such as those found at MIT and ASU, arose from the integration of STEM content into established fields in the humanities and arts. Nevertheless, the committee struggled to find evidence of student learning outcomes associated with this form of integration.

Several possible reasons could account for the lack of published course evaluations of STEM integration into the arts and humanities relative to arts and humanities integration into STEM. One hypothesis is that the culture of STEM encourages more data collection and publication in peer-reviewed journals. In this case, faculty based in a STEM discipline might be more likely to carry out an evaluation of student outcomes from an integrative educational experience and publish it in a journal. In contrast, the culture and scholarship of the humanities and arts lends itself more to argument-based essays, in the case of the humanities, and exhibitions and demonstrations, in the case of the arts.

The committee also found abundant evidence of a growing interest and demand for integration. Faculty members teaching integrative courses and programs who spoke to the committee expressed great conviction that the integrative model has benefited their students. This sentiment is also captured in the responses the committee received to its "Dear Colleague" letter, along with the examples that can be seen in the next section, entitled "Gallery of Illuminating and Inspirational Integrative Practices in Higher Education." Moreover, the committee catalogued 218 examples of integrative courses and programs at a range of institution types. While this is not an exhaustive sample, and it undoubtedly overlooks many programs and courses, the numbers point to the fact that there is buy-in for this approach at many different institutions and institution types (see "Compendium of Programs and Courses That Integrate the Humanities, Arts, and STEMM," available at https://www.nap.edu/catalog/24988 under the Resources tab).

Gallery of Illuminating and Inspirational Integrative Practices in Higher Education

The following cases (and other examples embedded throughout this report) are representative samples of a wide range of possible outcomes from integrative artistic practices that are connected to higher education, including programs, courses, and collaborations. Although not an exhaustive list, they reflect diverse approaches and applications to integration, and the various modes and pathways that different disciplines can use to inform one another to produce a myriad of results, outcomes, synergistic impacts, and localized, as well as far-reaching, potential benefits. Examples include a range of research projects, performative- and exhibition-based work, curricular and co-curricular endeavors, and community-engaged projects. Categories include citizen science, science communication and engaged research; therapeutic interventions and storytelling to inform innovation, healing, and discovery; exhibition and installation; venture creation; and aesthetic, as well as scientific explorations in the lab, with new media, and through traditional, as well as nontraditional, performance venue environments. These examples are from U.S.-based projects and are affiliated with academic institutions demonstrating curricular, co-curricular, and research-based integrative models and pathways (see Compendium of Programs and Courses on the Integration of Humanities, Arts, and STEMM at nap.edu). The committee maintains that the impact of certain forms of knowledge creation cannot be sufficiently described in words or numbers.

Some of these examples are drawn from work compiled by:

SEAD at https://seadexemplars.org/

SEAD steering group: Roger Malina, Carol Strohecker, Robert Thill, Nicola Triscott, Robert Root-Bernstein, Carol LaFayette, and Alex Garcia Topete.

And XSEAD at http://xsead.cmu.edu/

Thanssis Rikakis (Principal Investigator, Vice Provost for Design, Arts and Technology + School of Design), Aisling Kelliher (Senior Personnel, School of Design), and Daragh Byrne (Lead Developer, School of Design)

KINETROPE: CREATING CROSS-DISCIPLINARY SPACES TO PROMOTE DISCOVERIES AND CHANGED PERSPECTIVES

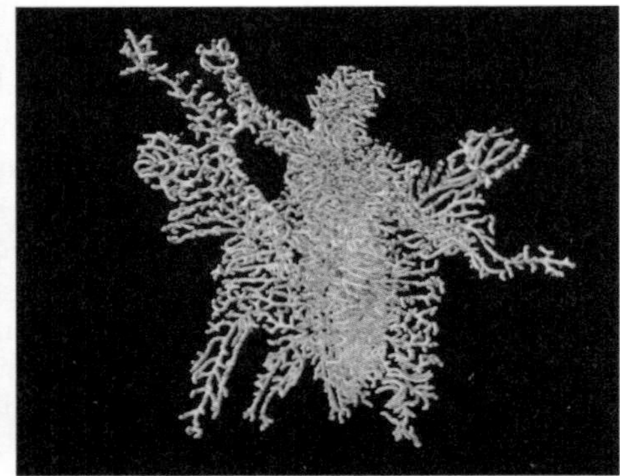

Project name: Kinetrope (2015)

Created by: Timothy Wood (Ph.D. Student, transLAB, Media Arts and Technology, University of California, Santa Barbara)

The transLAB (*The Transvergent Research Group*) is a research facility whose mission is to investigate how technology alters the relation between actual, transactivated, and virtual space in art and science by bringing together diverse expertise to investigate areas such as worldmaking, trans-modality, computational composition, algorithmic aesthetics, n-dimensional space, space as interface, and more. For example, Kinetrope is a plant-like life form born in virtual space that grows toward the motion of another. In reality there are many types of tropisms that give plants the impulse to grow and to be shaped in specific and unique ways. Through the creation of this specific growth algorithm and a new type of tropism, the possibilities of relationships and choreographies across boundaries becomes possible.

Timothy Wood (fishuyo@mat.ucsb.edu)

Project website: http://fishuyo.com/projects/kinetrope/transLAB website: http://translab.mat.ucsb.edu

CITIZEN SCIENCE/CITIZEN ARTIST AND COMMUNITY ENGAGEMENT

Crude Life Portable Museum: A Citizen Art and Science Investigation of Gulf of Mexico Biodiversity after the Deepwater Horizon Oil Spill

Brandon Ballengée (Louisiana State University); Prosanta Chakrabarty (Louisiana State University); Sean Owen Miller (University of Florida); and Rachel Mayeri (Harvey Mudd College)

Crude Life is an interdisciplinary art, science, and outreach project focused on gathering data on endemic fishes affected by the 2010 Gulf of Mexico Oil Spill. This engagement project raises public awareness of local species, ecosystems, and regional environmental challenges through community "citizen science" surveys and a portable art-science museum of Gulf of Mexico and Acadiana regional biodiversity. A project focus is engaging area fishers and other community members to look for 14 missing species of endemic Gulf fishes that have not been found following the spill.

PERFORMANCE AS PLATFORM FOR BUILDING BRIDGES BETWEEN DISCIPLINES

Hypermusic Prologue: A Projective Opera in Seven Planes

Lisa Randall (Harvard University) and Hector Parra (composer)

SOURCE: https://seadexemplars.org/portfolio_page/hypermusic/

Hypermusic is a project that demonstrates how music and storytelling can come together to facilitate a better understanding of science. The multi-dimensional opera brings together music, performances, and story to represent a new model of space-time based on contemporary research and theory. This opera was designed to disseminate the basic principles of Professor Lisa Randall's space-time model among the physics community and the artistic world. According to the SEAD website, "the intent of Hypermusic was to make a new scientific concept—a new space-time model—more understandable for fellow physicists and other scientists while also experimenting with the music and storytelling."

IMAGINING A BETTER FUTURE THROUGH CREATIVE WRITING

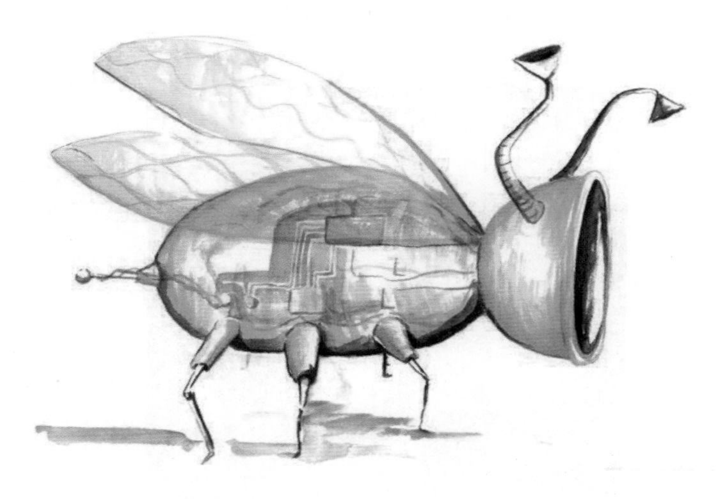

Hieroglyph: Stories and Visions for a Better Future

Illustration for the story "Johnny Appledrone vs. the FAA," in *Hieroglyph: Stories and Visions for a Better Future*. Edited by Ed Finn (Arizona State University) and Kathryn Cramer (writer).

Illustration by: Haylee Bolinger (https://www.hayleebolinger.com)

SOURCE: https://seadexemplars.org/portfolio_page/hieroglyph/

Hieroglyph is a project of Arizona State University's Center for Science and the Imagination that curates the innovative work of writers and researchers, bringing together top science fiction authors with scientists, engineers, and other experts to collaborate on futuristic visions grounded in real insights from science, technology, and a wide range of other disciplines. According to Project Hieroglyph's website (http://hieroglyph.asu.edu), "certain iconic inventions in science fiction stories serve as modern *hieroglyphs.*" Hieroglyph's first anthology, *Hieroglyph: Stories and Visions for a Better Future*, features 17 original stories born from these collaborations, demonstrating how art and storytelling can rekindle our grand ambitions for the future. The book includes innovative proposals such as 3D printing in space, an alternative internet powered by drones, solar cities designed to mimic algae cells, and more. The book sparked a national conversation throughout news outlets about the role that science fiction plays in igniting the public's imagination and bridging our present with the future. This project has served as a model for futurism and its power to shape innovation.

CULTURAL DISPLAY OF THE INTEGRATION OF ART AND SCIENCE

Blue Morph

Victoria Vesna (University of California, Los Angles); James K. Gimzewski (University of California, Los Angles)

SOURCE: http://artsci.ucla.edu/BlueMorph

BLUE MORPH is an interactive installation that uses nanoscale images and sounds derived from the metamorphosis of a caterpillar into a butterfly. According to the artist's website, "Nanotechnology is changing our perception of life and this is symbolic in the Blue Morpho butterfly with the optics involved—that beautiful blue color is not pigment at all but patterns and structure which is what nano-photonics is centered on studying. The lamellate structure of their wing scales has been studied as a model in the development of fabrics, dye-free paints, and anti-counterfeit technology such as that used in monetary currency." Victoria Vesna is the Director of the Art|Sci center at the School of the Arts and California Nanosystems Institute (CNSI). James Gimzewski is a Distinguished Professor of Chemistry at the University of California, Los Angeles and Director of the Nano & Pico Characterization Core Facility of the California NanoSystems Institute.

CREATION OF SOLUTIONS THAT IMPROVE LIVES AND CREATE NEW INDUSTRY MODELS

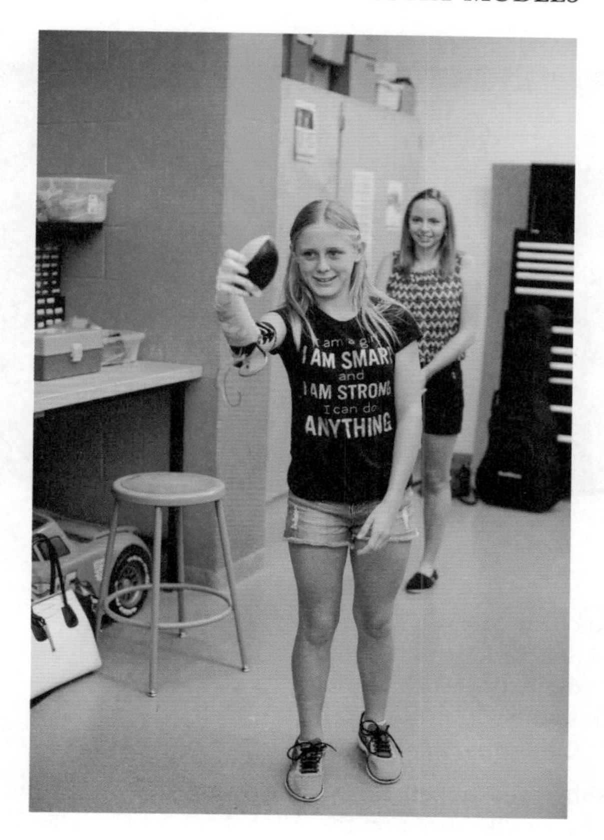

Limbitless Solutions

Albert Manero, President (University of Central Florida)

Annika Emmert using Limbitless Solutions' flower arm.

Limbitless Solutions is a nonprofit organization that uses 3D printing to create personalized bionics and affordable prosthetics. The organization grew out of work that the founding team members did as students at the University of Central Florida. According to group's website, the organization is "devoted to building a generation of innovators who use their skills and passion to improve the world around them. . . . We believe that no family should have to pay for their child to receive an arm. Now we want to lead by example and encourage communities to innovate with compassion."

CREATING RESEARCH FACILITIES THAT ARE IMMERSIVE AND TRANSDISCIPLINARY

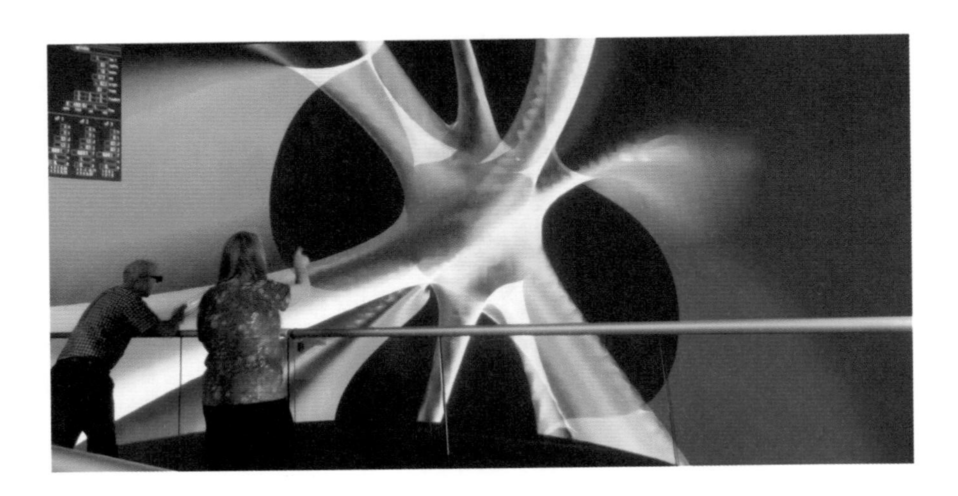

AlloSphere

Dr. JoAnn Kuchera-Morin
Director, AlloSphere Research Group
Professor, Media Arts and Technology
Professor, Music
University of California, Santa Barbara

This image shows researchers controlling parameters of real-time visualization of a hydrogen-like atom on the AlloSphere bridge. The AlloSphere, a 30-foot diameter sphere built inside a three-story near-to-anechoic (echo free) cube, facilitates research collaborations in an environment that can simulate reality. The AlloSphere allows use of multiple modalities to represent large and complex data, incluing immersive visualization, sonification, and interactivity. According to the AloSphere website, "We are creating technology that will enable experts to use their intuition and experience to examine and interact with complex data to identify patterns, suggest and test theories in an integrated loop of discovery. Important research areas include quantum information processing and structural materials discovery, bioengineering and biogenerative applications, and arts and entertainment." Dr. Kuchera-Morin serves as the Director of the AlloSphere Research Facility at the University of California, Santa Barbara.

The IndaPlant Project: An Act of Trans-Species Giving

Elizabeth Demaray (Rutgers); Qingze Zou (Rutgers); Simeon Kotchoni (Rutgers); and Ahmed Elgammal (Rutgers)

SOURCE: https://seadexemplars.org/portfolio_page/indaplant-project/

The IndaPlant Project is designed to facilitate the free movement and metabolic function of ordinary houseplants. It merges plants and robots in a way that creates an automated environment focused on the nurture of the plants. IndaPlant required adapting innovations from computer science and robotics in order to decode the plant-generated bio-information, and model solutions that allowed the plant-robots to seek sunlight and water. The floraborg (a term to describe an entity that is part plant and part robot) could allow for automated biodomes that would benefit plants and humans. Addressing the super sensory capacities of plants, this interface allows humans to decipher plant-based information on ecosystem health, the effects of climate change, and air pollution.

PIONEERING INTEGRATION AND THE CREATION OF NEW INDUSTRIES

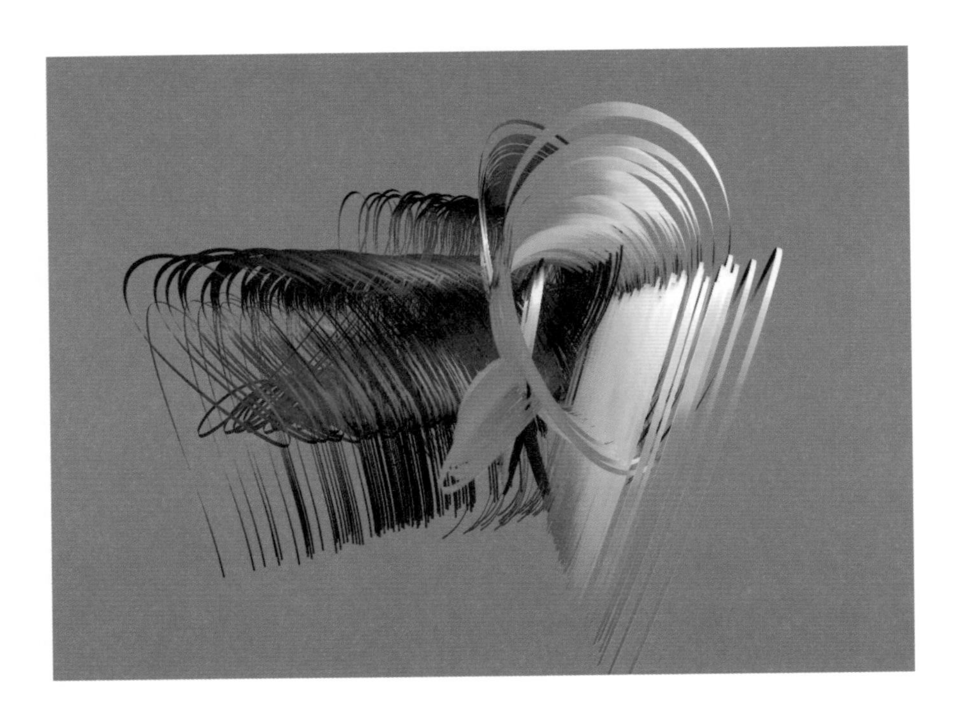

Ordo 211

Charles Csuri (Ohio State University)

Charles Csuri is an artist who has pioneered new technologies that have led to the creation of new industries. Csuri's pioneer research in computer graphics and animation has not only been applied to flight simulators, computer-aided design, visualization of scientific phenomena, magnetic resonance imaging, education for the deaf, architecture, and special effects for television and films, but also helped establish leading educational programs that trained the new professionals. His former students have worked for Industrial Light and Magic, Pacific Data Images, Metro Light, Pixar, Rezn8, Silicon Graphics Inc., USA Today, Rhythm and Hues, Xaos, Walt Disney Productions, and others.

Gemini (Permanent Collection at SFMOMA)

By Neri Oxman (MIT Media Lab)

In collaboration with Stratasys and Prof. W. Craig Carter (Department of Materials Science and Engineering, MIT)

Paris, 2014

Photo: Michel Figuet

The design of Gemini—an acoustical "twin chaise"—"includes a number of length scales ranging from structure to material composition that affect its sound absorbing properties: (1) On the meter scale, the chaise forms a semi-closed anechoic-like chamber with curved surfaces that tend to reflect sound inward. The surface structure scatters the sound and absorbs it and, in the absence of large planar surfaces, reduces the amount of sound that would otherwise bounce back to the source; (2) On the centimeter scale—a scale that corresponds to the wavelength of sound—the 3D printed inner 'skin' is designed as 3-dimentional doubly curved cells that scatter and absorb sound effectively given their geometry (i.e. the sound tends to bounce from one 'cell' unit to another till it gets absorbed) and high surface area to volume ratios. The features of the chaise are on the order of the wavelength of sound and they therefore interact strongly with sound and get absorbed effectively; (3) On the nano-scale, the properties of the Digital Materials also contribute to the absorption of sound. These materials are elastic in nature, varying in durometer (and sound absorption) as a function of curvature. Surface areas that are more curved than others are also assigned more elastic properties, thereby increasing absorption around local chambers," according to Oxman's Material Ecology website (http://www.materialecology.com/).

Ape Cinema

Rachel Mayeri (Harvey Mudd College)

SOURCE: https://seadexemplars.org/portfolio_page/ape-cinema/

Ape Cinema utilizes how common human practices can become means to explore other fields. The project consists of an original movie made expressly for a chimpanzee audience, who seem to be watching the same things as human primates: dramas around food, territory, social status, and sex. The project creates a prism for human beings to learn more about the complex social, cognitive, and emotional lives of chimpanzees by watching a movie through chimps' eyes.

ANIMATING RESEARCH AND ACTIVATING SPACES OF KNOWLEDGE PRODUCTION

As part of Liz Lerman's Animating Research class, students lead a participatory moment during a performance at the Biodesign Institute at Arizona State University (2017)

Credit: Deanna Dent/ASU Now

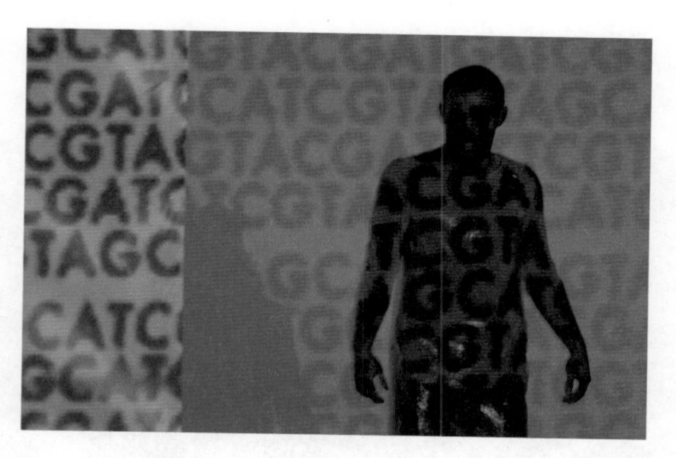

Performer Matt Mahaney in Liz Lerman's evening-length work Ferocious Beauty: Genome (2006).

Credit: Kevin Kennefick

Choreographer, MacArthur Fellow, and Institute Professor at Arizona State University (ASU), Liz Lerman created her "Animating Research" course to combine contemporary movement, dance, theater, and science into multimedia, immersive experiences for audiences and performers: a dozen artists were paired with molecular virologists, evolutionary biologists, and engineers to create mini-performances using and exploring ASU's biodesign building space. Lerman developed her process over decades of creative research through multidisciplinary works including her "science trilogy": Healing Wars (2014); The Matter of Origins (2011, with support from the National Science Foundation); and Ferocious Beauty: Genome (2006), created in collaboration with dozens of geneticists and educators, exploring how knowledge of the genome changes the way we think about aging, perfection, ancestry, and evolution.

Kirigami Inspired Elastic Solar Cells Model

Image of dynamic kirigami structure capable of solar tracking, consisting of monolithically integrated, single crystalline yet flexible, gallium arsenide solar cells on polyimide sheets. Work credit: Aaron Lamoureux, Kyusang Lee, Matthew Shlian, Stephen Forrest, Max Shtein—University of Michigan
Image credit: Aaron Lamoureux—University of Michigan

Matthew Shlian is an artist and designer working in paper using the traditions of origami, kirigami, and paper engineering to transform flat materials into 3D sculptures. Shlian, in collaboration with Max Shtein, material scientist and engineer at the University of Michigan, Ann Arbor, began exploring the question, "Can we make solar cells more efficient by using paper folding and cutting techniques?" The limitation of stationary solar panels is that they are only at maximum efficiency for a limited amount of time during the day. Normal solar tracking mechanisms are prohibitively bulky, heavy, and expensive for use on rooftops. Working together with the Stephen Forrest Group (an electrical engineer, also at the University of Michigan) and graduate students Aaron Lamoureux and Kyusang Lee, they developed solar cells made from thin-film crystalline gallium arsenide bonded to thin foils. The foils were then laser-cut into patterns inspired by kirigami, which allowed them to adjust the angle of the solar cells simply by stretching the kirigami sheet. As a result, exposure to the sun is maximized and electricity production is boosted significantly.

Tornado Project

The Cube

Bill Carstensen, David Carroll, Drew Ellis, Peter Sforza

Virginia Tech

Utilizing the technology made available at The Cube at Virginia Tech, the team created a 3D meteorological immersive experience of a tornado. According to the website, “The full scientific potential of radar data is not normally realized simply because current radar visualization is rather basic —typically on a flat screen on which, at best, static 3-D representations are rendered.” The Cube is a four-story-high, state-of-the-art theatre and high-tech laboratory that serves multiple platforms of creative practice by faculty, students, and national and international guest artists and researchers. The Cube is a highly adaptable space for research and experimentation in big data exploration, immersive environments, intimate performances, audio and visual installations, and experiential investigations of all types. This facility is shared between ICAT and the Center for the Arts at Virginia Tech.

CROSS-DISCIPLINARY COLLABORATION FOR COMMUNITY IMPACT

The Rain Project

Changwoo Ahn (George Mason University) EcoScience+Art

Photo by Evan Cantwell/Creative Services/George Mason University

SOURCE: https://www.changwooahn.com/ecoscience--art

EcoScience+Art is an initiative and collaboration between the arts and sciences at George Mason University with a mission to bring together individuals working across the boundaries of ecosystem science, art, and design fields to share knowledge, expertise, and strategies for creatively engaging in the common pursuit of a sustainable future. The Rain Project is a floating treatment wetland on Mason Pond. Students participate in a project-based learning approach aimed at developing innovative interdisciplinary education and scholarship. The goal is to raise awareness of critical stormwater issues for the Mason community, by means of a year-long project (Fall 2014 through Fall 2015) in which science collaborates with engineering, arts, and humanities in order to design and implement a floating wetland in Mason Pond.

The Very Loud Chamber Orchestra of Endangered Species

Pinar Yoldas

SOURCE: https://stamps.umich.edu/creative-work/stories/pinar-yoldas

Pinar Yoldas is a Turkish artist and scholar who works with the medium of speculative biology. In *The Very Loud Chamber Orchestra of Endangered Species*, she addresses the impact of anthropogenic forces on non-human animals. She takes skulls from endangered species and outfits them with tubes that push air through the skulls. In so doing, she works with the acoustics of their bone architecture, creating an orchestral arrangement of sounds. She writes, "By literally giving a VOICE to those whose habitats and lives are jeopardized by human activities, the project will initiate a subliminal emotional dialogue between viewers and the life forms that they often overlook. In essence, this project is an audible attempt to restore the dignity of other organisms that inhabit this planet and is an aesthetic amplifier of the negative consequences of our cultural choices. Alternatively, this project can be understood as a memento mori for those whose existence has been threatened, and a roaring wake-up call to the human race."

Ghost Food
Miriam Simun (MIT Media Lab) and Miriam Songster (Artist)
Images: courtesy of the artists

SOURCE: http://xsead.cmu.edu/works/115; http://www.miriamsimun.com/ghostfood/

Ghost Food is a mobile food truck that puts together sense and food pairings using a wearable device that helps engage smell. Scents of foods threatened by climate change are paired with foods made from climate change-resilient foodstuffs, to provide the taste illusions of foods that may soon no longer be available. Ghost Food staff serve the public, guiding visitors through this pre-nostalgic experience and engaging dialogue. Ghost Food has been deployed in Newark and Philadelphia thus far.

EXPLORING SOCIAL DISCRIMINATION THROUGH INTERACTIVE NARRATIVE AND GAMING TECHNOLOGY

Mimesis

D. Fox Harrell (Massachusetts Institute of Technology)

Video game technology and the creation of educational programs that utilize it is on the rise. This work by D. Fox Harrell at MIT is created to explore the idea that social networks and video game components can be used to help people better understand and create empathy, deliver meaningful experiences and enable critical reflection on identity. According to the Mimesis website, "The story of Mimesis takes place in an underwater setting with subtly anthropomorphized sea creatures as characters. The player character is a mimic octopus, which is a species of octopus adept at emulating other creatures. The octopus is on a journey that takes it from the dark depths of the ocean to its home in the tropical shallows. Along her way, the octopus will encounter several sea creatures who inhabit the waters and whose actions serve as examples of particular kinds of covert discrimination. These sea creatures provoke the octopus, leaving the player to must choose between different emotional responses to the creatures in order to guide the octopus through a series of short conversations. In this way, the project maps the experience of discrimination onto gameworld based on an underwater metaphor."

COMMUNITY, SKATEBOARDS, AND MOTION DATA CAPTURE

Gallaudet CRATERs: Finding a Line
Max Kazemzadeh, Dave Mutarelli, Ben Ashworth, Garth Ross, Dr. Dave Snyder
Gallaudet University and the Kennedy Center for Performing Arts
Photo credit: Max Kazemzadeh (Top); John Falls (Bottom)

CRATERS stands for Collaboration Research, Art Technology, Engineering, Robitics and Skateboarding. The Gallaudet CRATERs: Finding a Line Bowl was initially built by Dave Mutarelli (Lifetime Decks LLC) and Ben Ashworth (Sculpture Faculty at George Mason University) as commissioned by the Kennedy Center for Performing Arts. The Kennedy Center's "Finding a Line" event was led by its VP for Community Engagement, Garth Ross, who hosted a number of professional skateboarders, musicians, artists, and the public to skate and create together in the Center's front yard. After the event, the Kennedy Center donated the bowl to Gallaudet University in order to serve the creative, academic, and collaborative initiatives that began with a Special Topics course designed and taught by Gallaudet Associate Professor Max Kazemzadeh's titled "Skateboarding, Tracking & Data Visualization." The integrative course is supported by a NASA Space Grant initiated by Gallaudet Chemistry & Physics Professor Dr. Dave Snyder.

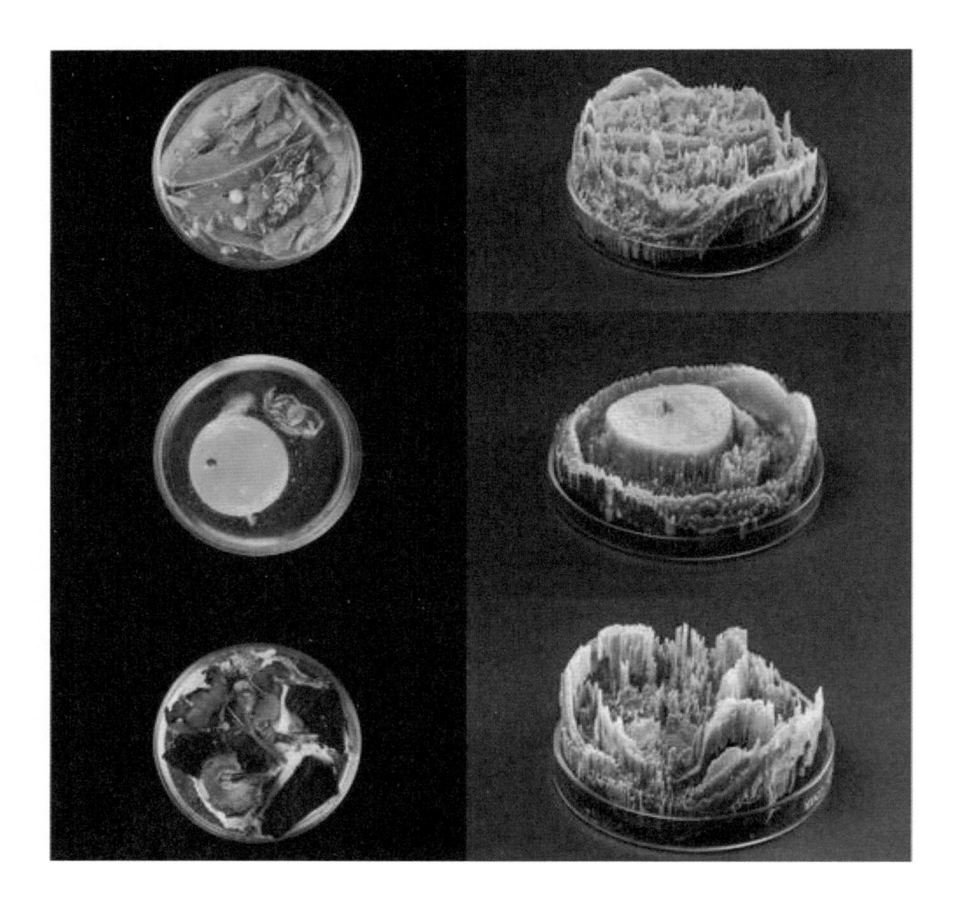

Vanitas (in a Petri dish)/Remote Sensing
Suzanne Anker (School of Visual Arts, NYC)

The Bio Art Lab is a fully functioning science lab housed within an art school. It was founded in 2011 as part of the School of Visual Arts' BFA Fine Arts facility in Chelsea, NYC. The Lab was founded and is directed by Suzanne Anker, Chair of the BFA Fine Arts Department. Conceived as a place where scientific tools and techniques become tools and techniques in art practice, the Lab provides a space for artists to investigate aspects of the biological sciences such as evolution, artificial life, and robotics through digital sculpture and new media installations.

7

Integration in Graduate and Medical Education[1]

Preparing the next generation of graduate students to tackle the complex problems facing the twenty-first century may necessitate a shift in thinking about graduate training (Begg et al., 2015; Begg and Vaughan, 2011; Borrego and Newswander, 2010; Bullough, 2006; Graybill et al., 2006). The argument is that a traditional, disciplinary approach to graduate education may not equip students with the awareness, knowledge, and skills needed to approach, frame, and solve contemporary and increasingly complicated problems, "which tend to defy traditional, disciplinary, and institutional boundaries" (Meyer et al., 2016, p. 1). Practicing scientists, as well as policy makers, have recognized the need for this shift toward interdisciplinarity in graduate research and education since the mid-1990s. For example, the National Academies of Sciences, Engineering, and Medicine (Institute of Medicine, National Academy of Sciences, and National Academy of Engineering, 1995) published the report *Reshaping the Graduate Education of Scientists and Engineers*, in which it recommends the development of a modified model for Ph.D. training that prepares graduate students for "an increasingly interdisciplinary, collaborative, and global job market" (Martin and Umberger, 2003, p. 87).

There have been similar calls for greater integration in medical training. In its 2017 Standards for Accreditation for medical schools, the Liaison

[1] The Committee wants to acknowledge and thank the research consultant, Dr. Matthew Mayhew, for his significant contributions to this chapter. A commissioned literature review written by Dr. Mayhew on behalf of the committee contributed directly to the writing of this chapter.

Committee on Medical Education, which is the accrediting body recognized by the Department of Education for all M.D.-granting programs, stated that medical schools should integrate social and behavioral sciences, societal problems, medical ethics, cultural competency and health disparities, communication, and interprofessionalism.[2] Many medical schools now offer required and elective courses that integrate the humanities and arts with medicine. The goals of such programs are varied but include the following: (1) ingrain aspects of professionalism, empathy, and altruism; (2) enhance clinical communication and observation skills; (3) increase interprofessionalism and collaboration; and (4) decrease burnout and compassion fatigue.

In this chapter, the committee reviews the literature on integrative and interdisciplinary graduate education and considers established fields of graduate study that are interdisciplinary. We also review the research on the positive learning outcomes associated with integration of the arts and humanities with medical training—a practice that has become widespread in medical education.

FEDERAL AGENCIES AND FOUNDATIONS HAVE SUPPORTED INTEGRATIVE GRADUATE EDUCATION

In response to the national call for graduate interdisciplinary education, the National Science Foundation (NSF) implemented the Integrative Graduate Education Research and Training (IGERT) initiative in 1998. This program challenged university faculty and administrators to develop interdisciplinary education programs to "meet the challenges of educating U.S. Ph.D. scientists and engineers who will pursue careers in research and education, with the interdisciplinary backgrounds, deep knowledge in chosen disciplines, and technical, professional and personal skills to become, in their own careers, leaders and creative agents for change" (Brown and Giordan, 2008, p. 2). In the first 10 years of this program, 4,232 doctoral students were exposed to interdisciplinary education and research through one of 195 grants at one of 96 American universities (Brown and Giordan, 2008). NSF has also supported interdisciplinary research through its CreativeIT program.[3] This program supported research in four main areas: (1) new theoretical models for understanding creative cognition and computation; (2) integrating creativity-based methods, practices, and theories to stimulate breakthroughs in science and engineering; (3) innovative educational approaches that encourage creativity; and (4) computational

---

[2] See https://www.aamc.org/members/osr/committees/48814/reports_lcme.html. (Accessed October 1, 2017).

[3] See https://www.nsf.gov/cise/funding/creativeit.jsp. (Accessed October 1, 2017).

tools and creative methods that support solution finding and problem solving (see Chapter 6 for additional information on this program).

Although the IGERT initiative, in particular, is one of the most well-known interdisciplinary training programs, other federal organizations have also designed similar strategies to foster interdisciplinarity among graduate students. The National Institutes of Health (NIH) started supporting the Clinical and Translational Science Awards (CTSA) program in 2006 to "advance integrated and interdisciplinary approaches to education and career development in clinical and translational science" (Begg et al., 2014, p. 2). This program emphasizes the role of interdisciplinary courses in promoting innovation when combined with discipline-specific education (Meyers et al., 2012). In a survey of CTSA education leaders, 86 percent felt that interdisciplinary team science training was important for graduate students, and 62 percent of those institutions offering interdisciplinary training did so in collaboration with another unit—such as a school of business, education, or law—within the university (Begg et al., 2014). The National Endowment for the Humanities has also supported interdisciplinary graduate work through a new program called the Next Generation Humanities PhD Planning Grants, which strives to prepare doctoral students more multidimensionally for the challenges facing academic institutions and the wider world.[4] In addition to the federal organizations, The Andrew W. Mellon Foundation has also supported integrative efforts through their Higher Education and Scholarship in the Humanities program. This program supports efforts at the graduate level that broaden the intellectual and professional preparation of students and "aspire to be transformative and integrative, rather than merely additive."[5]

These initiatives have increased the number of graduate students with interdisciplinary training, and evaluation of these programs has also contributed to a proliferation of literature on interdisciplinary and integrated learning within graduate education. Papers describing the results from the summative and formative assessments required by NSF and NIH, as well as the common elements of successful proposals, have added greatly to the scholarship on interdisciplinary training as both a process and an outcome. However, because most of these programs tend to focus on providing opportunities to integrate among similar disciplines, less is known about interdisciplinary education across fields, especially those initiatives looking to integrate aspects of the arts and humanities with science, technology, engineering, mathematics, and medicine (STEMM) graduate education,

[4] Personal communications with the National Endowment for the Humanities (Accessed October 1, 2017).

[5] See https://mellon.org/resources/shared-experiences-blog/revitalizing-graduate-education/) (Accessed October 1, 2017).

or vice versa. This type of integrated learning has gained more attention in recent years as a preferred practice in undergraduate education, as advocates champion its possibilities to enhance college student learning and development (Borrego and Newswander, 2010; Catterall, 2012; Dail, 2013; Ge et al., 2015; Grant and Patterson, 2016; Maeda, 2013; Sousa and Pilecki, 2013).

MOST OF THE RESEARCH HAS FOCUSED ON INTEGRATION IN GRADUATE EDUCATION BETWEEN SIMILAR DISCIPLINES

Despite the call for more integrated graduate practices that promote interdisciplinary work, and despite the existence of long-standing integrative graduate programs in established interdisciplinary fields (e.g., Science, Technology, and Society; Bioethics; Human–Computer Interaction; Gender and Ethnic Studies, etc.), such approaches to graduate education are represented in the literature, for the most part, across and between similar disciplines. The research literature suggests that scholars and researchers within the STEM fields, even those engaging in interdisciplinary work tend to work with other scholars in these fields. A similar within-disciplinary pattern is observed for educators in the humanities and the humanistic social sciences because these scholars tend to collaborate with peers in other humanities and social science fields, respectively (Bullough, 2006). Of course, this begs the questions: Are the interdisciplinary efforts in graduate education examined in the research literature effectively interdisciplinary? Are they effective at helping students integrate ideas across disciplines?

Several reasons have been offered to explain the challenges of interdisciplinary graduate practice. First, traditional and often static organizational structures within the academy, especially as they relate to graduate education, render interdisciplinary efforts (i.e., reconciling faculty loads between disciplines, credit-bearing differences between disciplines) logistically difficult to manage (Borrego and Newswander, 2010). Second, faculty often use their disciplinary frameworks for ascribing value to the interdisciplinary effort, with faculty in the humanities often framing successful integration as any student's progress toward achieving critical awareness (i.e., the critical, by definition, is the integrated) and those in the STEM fields viewing successful integration as collaborative work in teams (Borrego and Newswander, 2010; Engerman, 2015). These competing approaches to interdisciplinarity sometimes lead to conceptual confusions in course design, enacted practice, and, ultimately, student experiences. Finally, faculty within a discipline may not know how to approach problem solving from a discipline other than their own (Borrego and Newswander, 2010).

Integrative graduate work at the interface of the arts, humanities, and STEM subjects may face challenges similar to those faced by within-STEM interdisciplinary efforts. In 2010, the NSF CreativeIT program—a project from the Computer, Information Science, and Engineering directorate—and the National Endowment for the Arts sponsored a series of workshops to explore the challenges and opportunities of interdisciplinary research and education at the interface of science, art, and technology. Many of the challenges noted above were captured in workshop proceedings for the event: *Strategies for Arts + Science + Technology Research: A Joint Meeting of the National Science Foundation and the National Endowment for the Arts*, held at the NSF Headquarters in Arlington, Virginia. The goal of the workshop was to bring together a community-of-interest with unique interdisciplinary methods, requirements, and concerns. As captured in the StoryMap that resulted from this workshop, participants identified several challenges (see Box 7-1). The challenges noted by the participants included the following:

- "There are real and perceived differences in how we (artists, scientists, technologists) validate what we value."
- "Silos and unleveled playing fields create disparities in resources, infrastructure, and teaching-to-research ratios between disciplines."
- "Demonstrating impact of Art + Science + Technology (AST) research is difficult as research archives are not linked."
- "AST networks in the US tend to be part of academic clusters. They are vibrant yet closed to those outside of the system."

Our review of the published literature also revealed that a disproportionate number of articles on integrative graduate study framed interdisciplinary graduate work as a proxy for participation in interdisciplinary courses (Meyer et al., 2016; Newswander and Borrego, 2009; Posselt et al., 2017) or on interdisciplinary research teams (Borrego and Cutler, 2010; Borrego and Newswander, 2010; Hackett and Rhoten, 2009; Newswander and Borrego, 2009; Rhoten et al., 2009). As a result, and due to the lack of rigorous research designs, many of the lessons extracted from the empirical articles on interdisciplinary work reflect an emphasis on working on interdisciplinary teams, rather than on exposure to and participation in interdisciplinary graduate experiences.

Criticisms of this approach have questioned the use of work on interdisciplinary teams as the primary means for enacting interdisciplinary graduate practice: specifically, authors have cautioned that this emphasis may reinforce the mindset that "more (disciplines) is better" (Strengers, 2014, p. 550), overemphasize problem-based learning in interdisciplinary work (Stentoft, 2017), and neglect the nuances associated with how individuals solve interdisciplinary problems within interdisciplinary teams (Zhang and Shen, 2015). We caution readers in this regard: By centering interdisciplin-

BOX 7-1
StoryMap and Gap Analysis from Strategies for Arts + Science + Technology Research: A Joint Meeting of the National Science Foundation and the National Endowment for the Arts

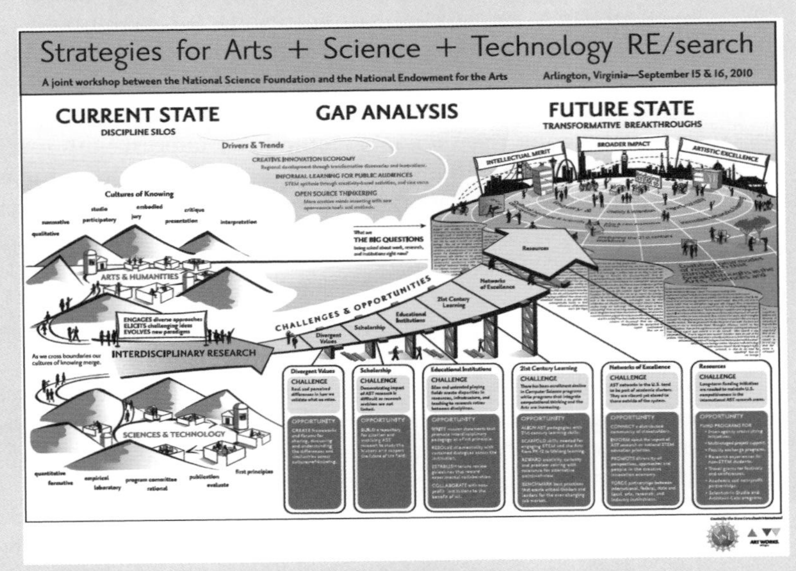

The current state landscape depicts the silos that often occur between disciplines clumped together in categories of Arts and Humanities and Sciences and Technology. There are a few emerging disciplines, institutions, and individuals that gravitate toward the ramp that connects the current and future state. The ramp represents the gap analysis topic areas. Above the ramp float drivers and trends that align the workshop topic with broader national and global opportunities in innovation, education, and economic development. Each gap analysis topic is represented by a ramp slab. Culled from a large quantity of workshop notes, each topic is divided into a "word-byte" for the "challenge" and the "opportunity."

The Future State world is perhaps better named "the future/now" as many of the goals, visions, missions, and activities are promoted in most progressive education, corporate, and nonprofit institutions. Champions and supporters of the goals of this gap analysis build the ramp pillars and assure its structural integrity to connect the "current" and "future" states. The "future state" is a land of transformative breakthroughs flanked by banners of the National Science Foundation and National Endowment for the Arts review criteria–Intellectual Merit; Broader Impact; and Artistic Excellence. The global city silhouettes emphasize that integrative research, practice and education strengthens the ties between our national and global communities. Jennings, P. & Shigekawa, J., Grove Consultants International (2010). Strategies for Arts + Science + Technology RE/search StoryMap, http://www.grove.com/cs_NationalScienceFoundation.php.

ary research teams as the hallmark of graduate interdisciplinarity, there may be an increased likelihood of assuming that working on these research teams will naturally produce the integration-based skills and competencies necessary for future work.

GRADUATE EDUCATION IN ESTABLISHED INTERDISCIPLINARY FIELDS

The foregoing discussion focuses on efforts to bring an integrative dimension to doctoral-level graduate training in traditional disciplines, but another important form of integration in graduate education is established programs in "interdisciplines," that is, in fields that have grown up at the intersection of established disciplines but have developed different approaches and areas of study than the disciplines that they grew out of. Such fields often focus on areas of intellectual and social importance that are neglected within the traditional disciplines. In this regard they are fundamentally different from the interdisciplinary initiatives discussed in the previous section. Examples of fields that include integrative approaches include Bioethics; Arts, Design, Media and Technology programs; Science and Technology Studies (including history of science, technology, and medicine); Sustainability; Cognitive Science; Gender Studies; Ethnic Studies; Geography; Environmental Humanities; Digital Humanities; and Medical Humanities. While some of these fields are young and in the process of taking shape, others are more well established and have their own scholarly approaches, professional societies, academic journals, and graduate training programs that are distinct from the disciplines they originally grew out of.

Yet all of these fields, in one way or another, integrate knowledge from STEMM, the arts, and the humanities. Although some are more science focused and some more humanities or arts focused, and although they may focus on very different points of intersection between the arts, humanities, and STEMM fields, they all draw upon and synthesize knowledge from various disciplinary domains to frame and address important questions in their fields, and in society more generally. Because they are sustained fields of inquiry, as opposed to relatively small-scale, one-off efforts at integration such as the IGERT programs discussed earlier, they offer importantly different opportunities and insights.

The graduate training programs that exist in these fields include important examples of programmatic approaches to integration at the graduate level. While there do not appear to be studies in the educational literature that systematically evaluate learning outcomes associated with these types of graduate programs, the growth of fields such as Sustainability; Science, Technology, and Society Studies; and Women's and Gender Studies reflect both the interest of prospective graduate students in training in integrative fields and a recognition by certain institutions of the importance of such programs within the portfolio of graduate education. Thus graduate training programs of integrative interdisciplines warrant attention, both as well-established spaces for doing programmatic integrative education and as sites for further attention and study.

Some of these fields also exemplify integrative approaches to socially important issues and the sorts of grand challenges that are not adequately addressed within siloed disciplines (see introduction). For instance, the field of STS grew in part out of the recognition that the ever-increasing importance of science and technology in social life demanded sustained scholarly attention to their social, cultural, political, and ethical dimensions (Jasanoff, 2010). Women's Studies developed as a critical response to systemic power asymmetries between the sexes (Ginsberg, 2008). Sustainability, a younger but rapidly growing field, reflects a recognition that the environmental challenges of our day are multidimensional and require an integrated response that attends to social, cultural, economic, and political dimensions alongside scientific and technological ones (Miller et al., 2014b). Such fields have been important domains of innovation in integrative approaches in both scholarship and in graduate training. In addition, these fields are important reservoirs of expertise about how to do robust, integrative interdisciplinary scholarship and teaching. Students trained in these fields learn how to move between disciplinary approaches, recognize their critical limitations, and integrate them—skills that are invaluable for future teachers who will provide integrative learning opportunities at the undergraduate level.

Indeed, these fields can (and in some cases already do) play an important role in integrative, undergraduate education. Graduate programs that train faculty in integrative scholarly approaches also produce future teachers. While there are certainly examples of innovative integrative teaching by faculty trained in traditional disciplines, the committee heard from experts who noted that high-quality integrative education benefits from well-qualified teachers whose own training bridges between disciplinary domains in an integrative way. Indeed, an increasing number of faculty trained in integrative graduate degrees such as Arts, Design, Media and Technology, Medical Humanities, Bioethics, and STS are providing integrative undergraduate education, including in some of the undergraduate courses and programs discussed in earlier chapters (see "Compendium of Programs and Courses That Integrate the Humanities, Arts, and STEMM").[6] Therefore, one important outcome of integrative graduate education can be to increase institutional capacity for providing integrative undergraduate education.

---

[6] Available at https://www.nap.edu/catalog/24988 under the Resources tab.

While undergraduate opportunities for integrative learning can be enhanced by integrative graduate programs, both depend on the presence of qualified faculty. This, in turn, requires institutional support for integrative scholarly fields and faculty, for instance, in faculty hiring and in support for research. Training future faculty will be an effective means to contribute to integrative learning primarily if there are faculty positions for graduates of interdisciplinary degree programs to occupy, and only if the promotion and tenure requirements associated with those positions recognize and reward integrative teaching and research. Traditional disciplinary control of faculty positions and of promotion and tenure requirements can be a significant barrier (see "The Disciplinary Segregation of Higher Education" in Chapter 2); however, because criteria of faculty evaluation are entirely within institutional control, criteria can be adjusted to encourage development of faculty whose research and teaching contribute to integration.

In short, integrative education at both the undergraduate and graduate levels is not independent of broader support for both research and training in integrative, interdisciplinary fields. Providing such support requires recognizing the value of investing in such programs, including by recognizing the contributions of established, integrative interdisciplinary fields where institutional infrastructure for supporting integrative education already exists and does not need to be built up from scratch, but simply supported, valued, and capitalized upon. If integrative education has a central role to play in twenty-first century higher education, then interdisciplinary fields that provide graduate training that integrates across arts and humanities and STEMM fields are likewise central to this mission. As such, they represent an important resource for future efforts to integrate STEMM, the arts, and humanities, both as fields capable of directly contributing to integrative education and as reservoirs of experience and expertise about interdisciplinary innovation.

THE IMPACT OF GRADUATE PROGRAMS THAT INTEGRATE THE HUMANITIES, ARTS, AND STEMM

Unfortunately, little has been published about the impact of graduate programs that integrate disciplines from the STEMM fields with those in the arts and humanities. This is surprising, given the number of graduate programs that integrate the humanities, arts, and STEM subjects (see "Compendium of Programs and Courses That Integrate the Humanities, Arts, and STEMM").

The committee faced two main challenges when reviewing the research on graduate education. First, there are few publications that describe graduate programs designed to integrate STEMM fields with those in the arts and humanities, and second, most of the programs described in the published literature did not provide ample information regarding the programs' efficacy in helping students achieve the desired learning outcomes. Nevertheless, the programs the committee considered describe positive outcomes and provide a sense for the motivations of such programs.

For example, at the University of Oklahoma, faculty developed the Designing for Open Innovation course as an interdisciplinary, flexible, learner-centric environment for engineering graduate students (Ifenthaler et al., 2015). This course, which is organized around addressing economical, sociological, and environmental dilemmas arising from energy policy, uses content-based lectures along with student teams to support learning. Researchers used measures of engineering attitudes, engineering self-concept, and team-related knowledge as well as written mental models to assess student learning. By comparing posttest to pretest scores, they found that students who participated in this course made gains in positive attitudes toward engineering, confidence in their ability to perform engineering tasks, and higher team-related knowledge.

Rider University created an integrative graduate program that combined global studies, environmental literacy, and corporate social responsibility into a set of interdisciplinary, study abroad "Business-Science" courses (Denbo, 2008, p. 215). These courses were designed to achieve the "dual goals of fostering integrated learning and global studies" (p. 215) and provided undergraduate science and liberal arts students, undergraduate business students, and graduate business students an opportunity to "not only study, but also experience, the complex interaction of legal, economic, social, and environmental factors in an international setting" (p. 216). Faculty specializing in marketing and legal studies worked with their colleagues in geological and marine sciences and biology to plan the course, which included both international science- and business-related site visits. Students participated in three courses (each lasting 4 hours) prior to the study abroad component. They learned about the legal, economic, and political structure of the country as well as sustainable business practices, environmentally sustainable marketing issues, corporate social responsibility, environmental law, and ecotourism related to their trip, as well as geological features and biodiversity of the destination country.

Upon arriving at the international sites, students met with local business representatives and toured their facilities (Denbo, 2008). They also engaged in scientifically focused excursions, visited a local university, and participated in cultural events. While abroad, students wrote daily in journals as a way to reflect on their experiences, and they completed a research paper or science project upon their return. Denbo (2008) provides some description of how faculty evaluated student learning through the assignments and indicated that students perceived the experience positively (Denbo, 2008).

Presenting quotes from MBA students as evidence for her claims of program interdisciplinary efficacy, she noted the following:

> Significantly, many of the students indicated that the trip was personally enriching and that they now feel more confident about their ability to travel or work abroad on their own. All students responded that the course broadened their understanding of the interaction of business and science, as well as the importance of being able to conduct business in different parts of the world. (pp. 236-237)

The nanomedicine program at Northeastern University, which provides a unique interdisciplinary opportunity for graduate students to apply nanoscale and nanotechnology to medical problems (van de Ven et al., 2014), offers an additional example. Though primarily focused within STEM, students in this program also learn how to "translate basic research to the development of marketable products, negotiate ethical and social issues related to nanomedicine, and develop a strong sense of community involvement within a global perspective" (p. 23) through programmatic elements, such as "experiential research, didactic learning, networking, and outreach" (p. 24). Program leaders developed four interdisciplinary specialized courses, a weekly seminar series, and outreach activities in K–12 public schools to complement the formalized dual mentoring that students receive from program faculty and placement in a required nonacademic internship. Students provide yearly progress reports to a faculty committee that evaluates their progress and ensures compliance with the program.

From 2006 to 2014, the Northeastern University nanomedicine program recruited 50 doctoral students in 10 different departments (e.g., bioengineering, biology, chemistry, electrical and computer engineering, mechanical and industrial engineering, materials science and engineering, and physics), 54 percent and 22 percent of whom identify as women and underrepresented racial minorities, respectively (van de Ven et al., 2014). These enrollment numbers exceed the national standards for IGERT programs (Carney et al., 2011, as cited in van de Ven et al., 2014), as does the nanomedicine program's retention rate of 94 percent. Van de Ven et al. also note that the program's trainees "have published 117 peer-reviewed manuscripts and presented at 189 conferences" (p. 26) as of 2014, with program Ph.D. graduates producing "an average of 3.9 manuscripts and 4.1 conference presentations directly related to their Nanomedicine project" (p. 26). These graduates have also pursued careers in the health care sector and academia or have started their own companies.

Other programs, such as the Research-oriented Social Environment (RoSE) project at the University of California (Chuk et al., 2012), have focused on the integration of the humanities and engineering. This socially networked system was designed to "represent knowledge in the form of relationships between people, documents, and groups" (p. 93) and was designed by a team of scholars in the humanities, software engineering, and information studies. This type of collaboration represented a central component of the emerging field of the digital humanities.

One of the challenges for the digital humanities noted by Chuk et al., (2012) was developing "collaborations that bring together the expertise of the various disciplines invested in it" (p. 94). For example, the scholars from the humanities, software engineering, and information studies were all essential to the development of the RoSE project. More specifically, humanities scholars brought critical inquiry to the discussion, suggesting improvements to RoSE and general reflections on the purpose and goals of the project. The information scientists introduced an organizational approach to metadata that mediated the distinction between real entities (persons) and their digital representations. The software engineers translated abstract ideas into formal models that resulted in the material form of the RoSE project as a system (p. 96).

However, these collaborations were not always easy and simple. Chuk and colleagues (2012) noted that the aims and epistemologies of the disparate disciplines often conflicted during work on the RoSE project. As a result, the authors recommend scholars on similar teams be aware of their disciplinary assumptions and be willing to vet unfamiliar new ideas with others. As graduate students working on this project, Chuk et al. (2012) concluded that the RoSE project provided an opportunity for synthesis of differing disciplinary perspectives and interrogation of disciplinary assumptions.

The University of Michigan has one of the only interdisciplinary graduate residency programs, the Munger Graduate Residencies.[7] They actively recruit students from all 19 academic units on campus and place them in living suites based on interests, not disciplines. Their website states, "Experience true multi-disciplinary collaboration. The world increasingly presents challenges that cut across multiple disciplines and skillsets. At the Munger Graduate Residences, a diverse mix of graduate and professional students from various fields live, study and interact together, building a culture of collaboration."

Among the many existing graduate programs that did not appear in the published literature are many that integrate art and technology. For example, graduate programs and degrees such as the New York University Integrated Technology program; the Carnegie Mellon University master of entertainment technology degree; the Stanford master of arts in music, science, and technology; the University of California Santa Barbara Media

[7] See http://mungerresidences.org/. (Accessed September 21, 2017.)

Arts and Technology program; the Dartmouth master in digital musics degree; the University of Miami master of science in music engineering technology degree; and the Georgia Tech M.S. and Ph.D. programs in music technology are only a few of the existing graduate programs that integrate music and STEM. It is unfortunate that more is not known about the impact of these programs on students.

WHAT CAN WE LEARN ABOUT PROMISING PRACTICES FOR INTEGRATION BETWEEN SIMILAR DISCIPLINES THAT COULD APPLY TO THE INTEGRATION OF THE HUMANITIES, ARTS, AND STEMM

Though the existing research literature cannot tell us much about the impact on students of graduate programs that integrate the humanities, arts, and STEMM, and the research on within-STEMM integration is also sparse, when taken together, some broad lessons can be extrapolated.

Some empirical efforts use innovative research questions to argue for the importance of interdisciplinarity in the graduate context. Also, some argue that interdisciplinarity may help graduate students from groups underrepresented in STEMM to thrive and negotiate the challenges of the graduate experience. Results also suggest that graduate interdisciplinary work should be supported at each stage of the graduate journey. As students, many interdisciplinarians struggle with finding their intellectual home, navigating priorities, and presenting scholarship (Calatrava Moreno and Danowitz, 2016; Graybill et al., 2006). As graduates, they must combat common notions within the academy that conducting interdisciplinary research hinders individuals' career advancement (Millar, 2013). In either case, educators who design interdisciplinary contexts for graduate students may need to adopt a developmental approach to supporting them—what may be perceived of and experienced as supportive by new graduate students may not be thus perceived and experienced by matriculating graduate students. Other studies concluded that discipline-specific content mastery should never be compromised for interdisciplinarity. Mastery of discipline-specific material was critical for effective engagement in graduate interdisciplinary practice.

Finally, the mechanisms for meaning making across disciplinary boundaries seemed to vary based on the task at hand. The idea that interdisciplinary thinking or ways of constructing information and solving problems is not naturally occurring appeared in several studies. This serves as an important reminder to educators that effective interdisciplinary practice needs intention and structure: Placing graduate students on interdisciplinary teams with the expectation that integration or interdisciplinarity will just occur is a fundamentally flawed practice.

SUMMARY OF EVIDENCE ON INTEGRATION IN GRADUATE EDUCATION

For the past 15 years, interdisciplinary training programs have emerged as a common practice for graduate education in the STEM fields, in particular. These initiatives combine cross-disciplinary curricula and courses, co-curricular seminars and workshops, and multidisciplinary research teams to prepare graduate students for the twenty-first–century workforce. Although collaboration among scholars in the social sciences, humanities and arts, natural sciences, and engineering exists in academia, the preponderance of interdisciplinary graduate programs described in the research literature occurs among those specializing in scientific fields, and to a lesser extent within the humanities or humanistic social sciences. These programs emphasize skill building as the primary focus of interdisciplinary work, and often use academic productivity metrics (e.g., counting interdisciplinary publications) and postgraduate career placements as indicators of success. Seemingly impressive, the lack of rigorous evaluation of these programmatic efforts (e.g., comparing program participants to nonparticipants) cannot conclusively indicate that program graduates achieve these interdisciplinary goals to any measured degree of difference than nonparticipants.

We know even less about interdisciplinary programs that combine the arts and humanities with the natural sciences and engineering. How are these programs developed? Which programmatic components do educators use and how are they different from "intra-STEMM" initiatives? And, most importantly, what learning outcomes do students achieve due to their exposure to and participation in these initiatives? Given what we know about integration in undergraduate education, it seems as though these types of interdisciplinary training programs have the potential to influence graduate interdisciplinary learning, yet the efforts described in the research literature do not currently provide us with enough information to support this claim.

Despite the limits of the existing research base on integrative graduate programs, there is much to learn from established integrative, graduate-level fields, such as STS, Gender Studies, Sustainability, and Bioethics, among others. These "interdisciplines" are an important resource for future efforts to integrate STEMM, the arts, and humanities, both as fields capable of directly contributing to integrative education at the graduate and undergraduate levels and as reservoirs of experience and expertise about interdisciplinary innovation.

INTEGRATION IN MEDICAL EDUCATION

Integrating the Arts and Humanities into Medicine

The integration of the arts and humanities into medical training is a widespread practice. Since the 1960s, the field of medical humanities has provided a framework and pedagogy for including the humanities in medicine. The goals of a medical humanities curriculum are to (1) ingrain aspects of professionalism, empathy, and altruism; (2) enhance clinical communication and observation skills; (3) increase interprofessionalism and collaboration; and (4) decrease burnout and compassion fatigue. The curriculum, which can include literature, poetry, narrative, theater, or visual arts as part of a medical education (Naghshineh et al., 2008), can be used as a way to help medical students develop their diagnostic skills or as a way to create more humanistic physicians (Wachtler et al., 2006). One result of this initiative has been the establishment of health humanities programs, which more than quadrupled from 2000 to 2016, going from 14 to 57 (Berry et al., 2016). For the 2015–2016 academic year, 94 percent of medical schools surveyed had required and/or elective courses in medical humanities (see Figure 7-1).

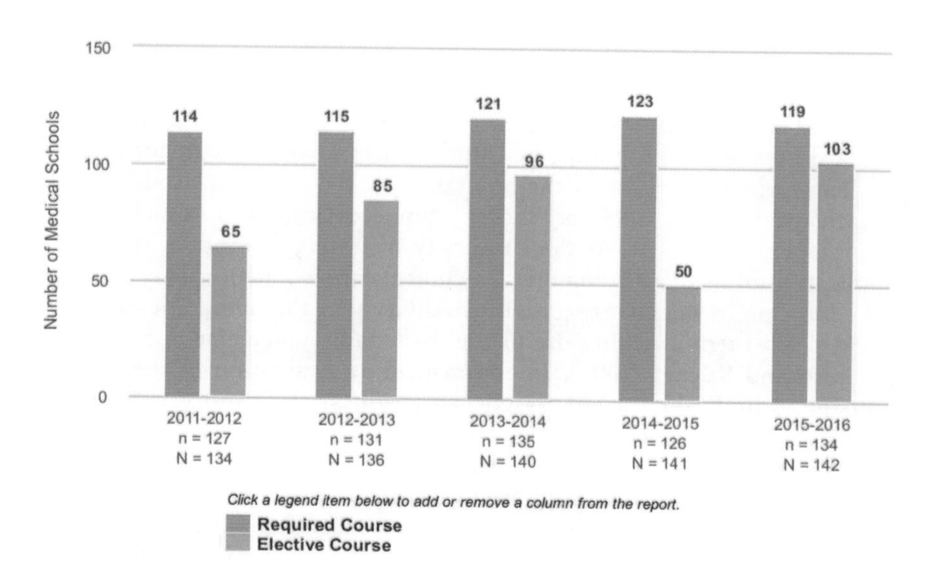

FIGURE 7-1 Data from the American Association of Medical Colleges showing the number of medical schools that include either required or elective courses in the medical humanities.
SOURCE: https://www.aamc.org/initiatives/cir/406462/06a.html.

A common method in the medical humanities is narrative medicine, which has been defined as "clinical practice fortified with a narrative competence to recognize, absorb, interpret, and honor the stories of self and others" (Miller et al., 2014). Narrative medicine creates a platform where the voices of patients and health care providers can be heard and valued in a way that produces a more humane way to practice medicine. Narrative medicine views patient histories as stories and analyzes them as one might unpack a novel's themes and plot strands. Using the intellectual tools and heuristics of close reading may enable students in a course in medicine, as Fraser (1987) observes, to liberate their thinking. Medical students who are given opportunities to perform artistically in groups or write poetry may learn to approach team learning more creatively. Studies of narrative medicine have demonstrated efficacy in increasing empathy, resilience, and teamwork (Sands et al., 2008). The medical humanities and other examples of integrating the humanities and art into medicine do not apply just to medical students and residents. Though much of the literature has focused on physicians-in-training, the entire health care team can be engaged in more humanistic medicine.

While medical humanities programs are widespread, curricula are not standardized across medical schools. Also, medical humanities programs typically require an interdisciplinary team of clinicians and instructors with backgrounds in the arts and humanities, creating a need for professional development and collaboration between multiple departments and institutions.

Further, perceptions of rigor are important for lending legitimacy to medical humanities programs. In a study of narrative medicine for medical students, when asked about the reputation of the program, students used negative stereotypes, such as "fluffy," "unnecessary," and "touchy-feely" (Arntfield et al., 2013). At the University of Irvine, California, where the application of medical humanities to clinical care is a major tenet of medical education, residents expressed incredulity that the humanities and arts could teach them anything useful for their daily management of patients (Shapiro and Rucker, 2003). In contrast, in an evaluation of the medical humanities and ethics scholarly concentration at the Stanford University School of Medicine, interviewees cited an enhanced sense of community, positive self-care and reflective practices, and rich and rigorous interactions with peers and faculty (Stanford Medicine, n.d.).[8] Many believed these skills to be necessary safeguards against physician burnout and moral erosion.

Despite skepticism among some students about the value of integrating the arts and humanities with medical education, accrediting bodies are

[8] More on the program is available at http://med.stanford.edu/bemh.html. (Accessed October 2, 2017).

now pushing for greater integration of medical education with a variety of other disciplines. In a 2008 report titled "Recommendations for Clinical Skills Curricula for Undergraduate Medical Education," the Association of American Medical Colleges included professionalism and communication skills in a list of competencies that all medical students and residents should acquire (AAMC, 2008). In its 2017 Standards for Accreditation for medical schools, the Liaison Committee on Medical Education (LCME, 2016), which is the accrediting body recognized by the Department of Education for all M.D.-granting programs, stated that medical schools should integrate social and behavioral sciences, societal problems, medical ethics, cultural competency and health disparities, communication, and interprofessionalism. Some efforts to adopt social medicine in medical school curricula have emphasized the interplay between biology and the social determinants of health (Westerhaus et al., 2015) or the integration of health systems science, which is an interdisciplinary field that draws from a variety of disciplines to improve health care delivery for both individuals and populations (Gonzalo et al., 2016). Eleven medical schools formed a consortium to implement a health systems sciences curriculum by identifying cross-cutting domains and curricular content (Gonzalo et al., 2017).

The integration of the arts in medical education is also a common practice that is often pursued with the goals of improving visual diagnostic skills, communication skills, and critical thinking among medical trainees. For instance, an elective course at Harvard Medical School for first-year medical students seeks to improve physical examination skills through a combination of observation exercises at the Boston Museum of Fine Arts paired with lectures on physical diagnosis (Naghshineh et al., 2008) (see Box 7-2). The observation exercises use a methodology called Visual Thinking Strategies to develop critical thinking, communication, and visual literacy. After taking the course, students were able to make more observations than a peer control group—for example, in describing patient dermatology photographs—and also used more fine arts terminology in their observations.

Art Rounds is another example of an innovative interdisciplinary program aimed at determining whether the use of fine arts instructional strategies benefits health professional education and is an elective course open to both nursing students and medical students at all levels. In their evaluation of the program, Klugman and Beckmann-Mendez note that "students were exposed to fine arts and taught to use Visual Thinking Strategies. The initial evaluation of the pilot program revealed improved physical observation skills, increased tolerance for ambiguity, and increased interest in communication skills" (Klugman and Beckmann-Mendez, 2015, p. 220).

The outcomes associated with the two preceding examples are consistent with those of several well-controlled studies that have demonstrated

BOX 7-2
Improving Physical Diagnosis Through Art Interactions

Training the Eye
OB/GYN residents from Brigham & Women's Hospital discuss a Kathe Kollwitz sculpture at the Museum of Fine Arts, Boston, using the Visual Thinking Strategies framework to make connections to their clinical practice.over 30 medical schools and museums across the country that collaborate in art and medicine learning activities.

that medical students, physicians, and nurses all benefit in a statistically significant manner from courses designed to educate visual observation skills through the examination and analysis of paintings and drawings (Braverman, 2011; Dolev et al., 2001; Grossman et al., 2014; Kirklin et al., 2007; Klugman et al., 2011; Naghshineh et al., 2008; Perry et al., 2011; Shapiro et al., 2006). Shapiro et al., (2006) found that medical students provided with 90-minute visual art and dance interventions weekly for 6 months had significantly improved pattern recognition skills compared with those viewing clinical photographs.

There are now many medical school–museum partnerships across the nation. The University of Texas at Dallas's Edith O'Donnell Institute for the Humanities hosted an Art Museum and Medical School Partnership symposium at the Museum of Modern Art in June 2016. The published report documented the attendance of 130 medical and museum partners in attendance. See Box 7-3 for an example from Virginia Commonwealth University.

Musical training also helps to develop pattern recognition and aural memory (i.e., aural imaging) alongside aural observation skills (Pellico et al., 2012; Sanderson et al., 2006; Wee and Sanderson, 2008). Mangione and Nieman (1997, 1999) tested 868 medical students and interns for their ability to learn how to distinguish between and correctly identify stethoscope recordings of 12 different typical heart diseases. Those who could play a musical instrument were statistically significantly more likely to get the diagnoses correct. Physicians and nurses also use aural observational skills when dealing with surgical and critical care equipment utilizing melodic alarm functions. Studies demonstrate that physicians and nurses who had previously played musical instruments were significantly better at discriminating between, correctly identifying, and responding to melodic medical equipment alarms used in surgery and critical care settings (Sanderson et al., 2006; Wee and Sanderson 2008).

This correlation is not limited to aural training among medical graduate students. Boyd et al. (2008) introduced a group of 30 medical students without previous training to laparoscopic surgical procedures and found that those with no music training learned the techniques slowest; those who had played an instrument at some time in the past but were not currently practicing learned the techniques faster; and those that currently played an instrument learned the techniques most quickly and efficiently. Harper et al. (2007) reached the same conclusion in their study of 242 medical students learning robotic suturing and knot-tying techniques. Students who had been athletes or musicians were very significantly ($p < 0.01$) able to learn robotic surgical techniques much more quickly, and to make fewer errors while doing so, than those without such training.

Another technique used to integrate the arts and humanities into medicine is anatomic bodypainting (Bennett, 2014). Anatomic bodypainting aims to teach by having students paint the underlying anatomy on a living and mobile subject. A student is able to gain an appreciation of performing a clinical examination by visualizing the manipulation of the painted structures, which cannot be achieved in the same manner by dissection.

One way in which medical schools have sought to further the integration of the arts and humanities into medical education is by actively recruiting humanities majors. For example, the HuMed program at the Icahn School of Medicine at Mt. Sinai, which began in 1987 and is now

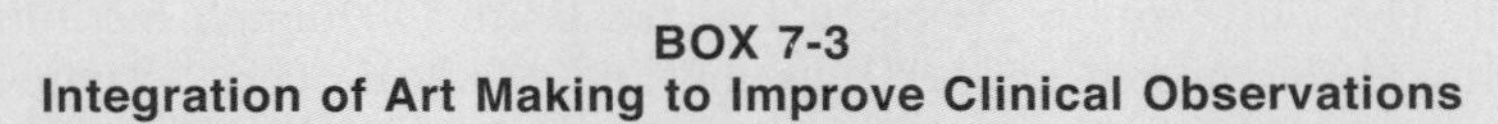

BOX 7-3
Integration of Art Making to Improve Clinical Observations

Plastic surgery residents participate in a figure sculpting workshop taught by 2011 Virginia Commonwealth University (VCU) sculpture alumna, Morgan Yacoe, in a partnership with the Center for Craniofacial Care at the Children's Hospital of Richmond at VCU directed by plastic surgeon Jennifer Rhodes. "Plastic surgery and sculpture have a great deal of crossover," says Rhodes, "Having Morgan willing to provide our division with a class here in Richmond is a fantastic opportunity for us to enrich our resident experience by encouraging them to think creatively, fostering an appreciation of subtle anatomic details and applying their anatomy knowledge in a different way."

SOURCE: https://www.news.vcu.edu/health/The_surgeon_as_sculptor.

known as the FlexMed program, offers early acceptance to sophomores with any undergraduate major and exempts them from taking the Medical College Admission Test. Students accepted through this alternative admissions process have not differed significantly from their traditional peers in academic achievement, such as clerkship honors, scholarly research, or first-author publications during medical school (Muller, 2014). Students from the HuMed program did have significantly lower Step 1 scores in the U.S. Medical Licensing Examination compared to their peers. Currently, the Icahn School of Medicine accepts about half of its medical school class through the FlexMed program.

One study of the Harvard "New Pathways" program, which began in the 1980s and incorporated psychosocial and humanistic concepts into the curriculum through problem-based learning, followed up with graduates 12 to 13 years after they matriculated at Harvard Medical School. Those who participated in the New Pathways program were more likely to be in primary care compared with a control group who did not partake in this curriculum, and they also rated themselves higher on a scale rating their preparation to practice humanistic medicine and ability to manage patients with psychosocial problems (Ousager and Johannessen, 2010).

A 2010 literature review of 245 articles about the medical humanities found that 224 of the articles praised the interventions or described and evaluated coursework (Ousager and Johannessen, 2010). While the 224 articles advocated for the inclusion of humanities coursework in medical education, only 9 of the articles sought to study the long-term impacts of medical humanities coursework. These articles examined the outcomes of integrating humanities in undergraduate medical education. One of the 9 papers referenced in the Ousager and Johannessen paper studied two groups of Harvard graduates, one from a traditional medical curriculum and one from a curriculum that was humanities oriented. Peters et al. found that students who came from the humanities-oriented curriculum were more prepared and inclined to pursue careers in humanistic medicine (such as primary care or psychiatry) compared to their peers who came from a more traditional medical curriculum. The students from the humanities-oriented curriculum were also more confident in managing patients' psychosocial issues (Peters et al., 2000). Another study by DiLalla et al. used a survey to assess medical students and practitioners and found that exposure to educational activities that involve empathy, spirituality, wellness, and tolerance correlates to an increase in empathy and wellness in medical practice (DiLalla et al., 2004). The literature review conducted by Ousager and Johannessen reveals that, while there is a wealth of course descriptions and advocacy of the medical humanities, there continues to be a shortage of studies reporting evidence.

INTEGRATING MEDICINE INTO THE ARTS AND HUMANITIES

Much of the literature on integration in medical education has focused on integrating the humanities and arts into medicine, but a few examples exist of medicine being integrated into the arts and humanities. Relatedly, arts therapy, which could be considered akin to arts *as* medicine, is being embraced by many who seek to use the restorative power of the arts to promote health and wellness. In this way, medicine acts as the inspiration for new applications of the arts.

One example of the integration of medicine and the arts is medical illustration (see Box 7-4). Medical illustration is a relatively small field, with only three accredited programs in the United States at Johns Hopkins University School of Medicine, University of Illinois at Chicago, and Augusta

BOX 7-4
Medical Illustration

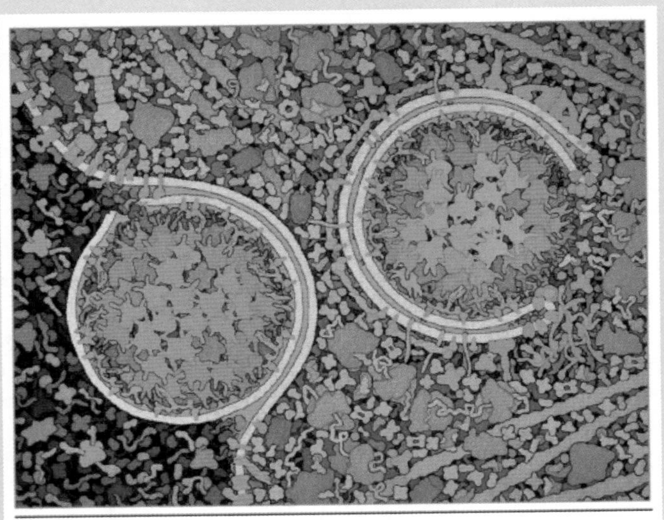

David Goodsell's watercolor painting of autophagy depicts the process by which a yeast cell packages waste material and delivers it to a vacuole. Courtesy of David S. Goodsell/Scripps Research Institute.

SOURCE: https://www.ami.org/medical-illustration/enter-the-profession/education/graduate-programs.

University. Medical illustrators use visual media to represent biological or medical information with an emphasis on communication. Medical illustration requires individuals to possess fine arts ability as well as strong biological and medical backgrounds (Bucher, 2016). Though there are few programs, it is worth noting the incredible impact of medical illustration on how emerging doctors come to understand and visualize the physiology of the human body and medical phenomena. Many medical phenomena are invisible to the human eye, but medical illustrators bring biological processes to life in striking definition (see Box 7-4).

Further, health, sickness, and death are themes that appear frequently in many media, such as visual art, literature, and theater. For example, graphic medicine is an area that depicts medicine using comics for both medical students and patients (Green and Myers, 2010). Deborah Aschheim, who has been an artist-in-residence at medical centers such as the University of California, San Francisco, is a visual artist who creates art installations using plastic, light, and electronics that reflect themes of memory and neural networks. In her artwork she tries to reconcile "the idea of the body as a kind of complex machine, with a more complex humanity" (UCSF Memory and Aging Center, 2011) (Box 7-5).

Creative arts therapy is used in the context of integrative medicine to address the physical, emotional, social, and spiritual dimensions of a patient in a holistic manner. There are well-established programs across the nation offering fully accredited therapeutic training in art, dance, music, and theater therapy. There are also numerous arts in medicine enrichment programs where artists-in-residence bring their craft to bear in a medical or therapeutic environment with patients. Some of the flagship programs that have distinguished themselves are the University of Florida's Center for Arts in Medicine, the University of Michigan's long-standing Gifts of Arts program, and the robust Performing Arts Medicine program at Houston Methodist Hospital. For example, one of the programs under the UCLArts and Healing Program, which promotes mental health through integration of the arts, is a group drumming intervention that has been shown to improve the social-emotional behavior of low-income children (Ho et al., 2011). Arts therapy is now being used to help veterans and military patients who have been diagnosed with traumatic brain injury and psychological health conditions (Lobban, 2014; Spiegel et al., 2006). Through the Creative Forces: NEA Military Healing Arts Network, the National Endowment for the Arts, the Departments of Defense and Veterans Affairs, and state arts agencies have partnered to bring together creative arts therapists, musicians, painters, potters, writers, woodworkers, dancers, doctors, military service members, veterans, community leaders, and policy makers to work to harness the "transformative and restorative powers of art" to "help military personnel and veterans return to their homes, their missions and their

BOX 7-5
The Work of Artist-in-Residence, Deborah Aschheim

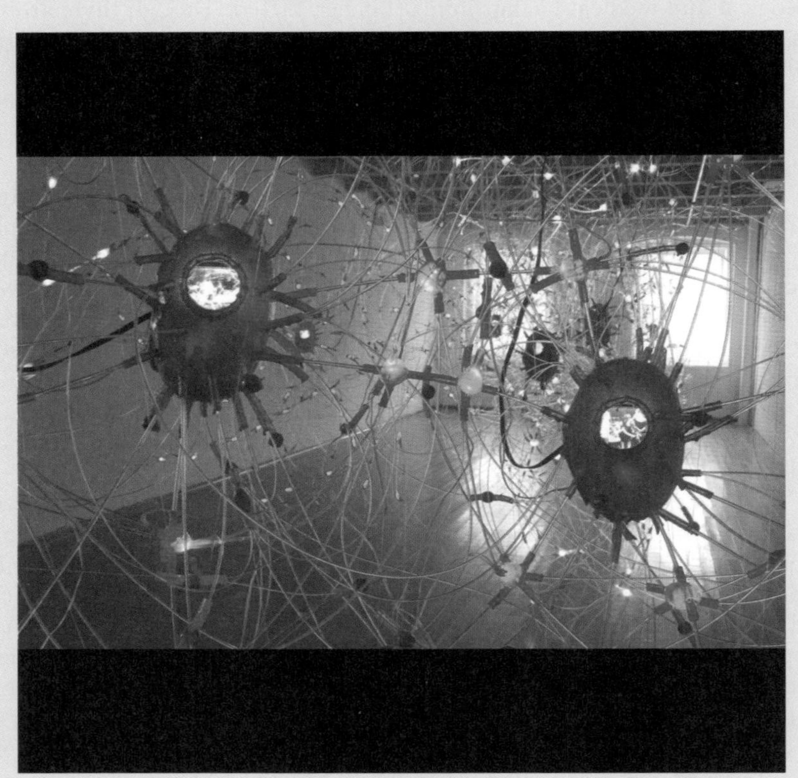

"On Memory" is a work by Deborah Aschheim (installation at the Mattress Factory, Pittsburgh, PA, 2006–7, plastic, video, LEDs.) This piece is one of many that explores themes of memory. Aschheim uses her work to "contend with the uncomfortable knowledge that systems and organisms invisible to us and often too minuscule to control, determine the quality and length of our lives. The phenomenon of memory belongs to one such physically minute but functionally far reaching realm" (Exhibition Review 2007).

families whole, mentally fit, and emotionally ready for whatever comes next[9] (Box 7-6).

Organizations such as Athletes and the Arts view performing artists as athletes and integrate concepts of sports medicine into training to optimize

[9] See https://www.arts.gov/partnerships/creative-forces.

BOX 7-6
Walter Reed Bethesda Art Therapy at the National Intrepid Center of Excellence

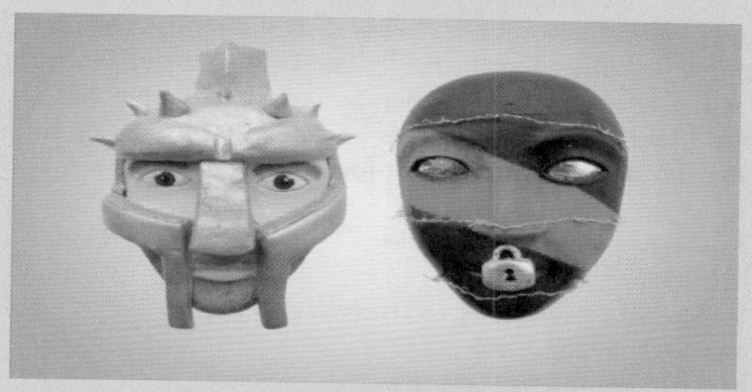

SOURCE: See https://www.ted.com/speakers/melissa_walker.

SOURCE: NICoE/Walter Reed Bethesda..

Melissa Walker is a Creative Arts Therapist at the National Intrepid Center of Excellence (NICoE), a directorate of Walter Reed National Military Medical Center, and Lead Art Therapist for Creative Forces: NEA Military Healing Arts Network. Walker works with active duty service members suffering from traumatic brain injury and psychological health conditions. Dedicated to helping recovering military service members safely express their deep thoughts and emotions in a creative environment, she designed the Healing Arts Program at NICoE in 2010 where she and her team engage patients in creative arts therapies—powerful treatments for helping them express their invisible wounds. Masks made by service members as part of the NICoE art therapy program were highlighted in a 2015 edition of National Geographic magazine. Walker received her Master's Degree in Art Therapy from New York University.

performance (Dick et al., 2013). This also creates recognition that performing artists are a unique and potentially medically underserved population.

An important and growing addition to this field in the past 15 years is arts and neuroscience, or neuroaesthetics. Johns Hopkins Medical Center opened a new International Arts + Mind Institute in January 2017 with the tag line "We're the source for cross-disciplinary discussion in brain science, architecture, music and art."[10] The University of Houston's School of Engineering was recently awarded a 5-year renewable National Science Foundation Ideas Lab grant for arts and neuroscience. The following are some of the subthemes and research foci:

- Music/Arts cognition studies
- Music/Arts and autism
- Music/Arts and stroke recovery
- Music/Arts cognition
- Music/Arts and Parkinson's
- Music/Arts and Alzheimer's
- Music/Arts and epilepsy

Other music and neuroscience programs and departments exist at universities throughout the country (see "Compendium of Programs and Courses That Integrate the Humanities, Arts, and STEMM" at https://nap.edu/catalog/24988 under the Resources tab). Such courses and programs combine music, psychology, acoustics, cognitive science, neuroscience, and evolutionary biology in varied and interesting ways (Society for Music Perception and Cognition, n.d.).

SUMMARY OF EVIDENCE ON INTEGRATION IN MEDICAL EDUCATION

Studies on the integration of the arts and humanities with medicine show a positive impact on students. Studies of narrative medicine have demonstrated efficacy in increasing empathy, resilience, and teamwork (Sands et al., 2008), while the integration of arts observation into medical training has been shown to improve visual diagnostic skills (Naghshineh et al., 2008), increase tolerance for ambiguity, and increase interest in communications skills (Klugman and Beckmann-Mendez, 2015). Students who participated in a medical school curriculum at Harvard that incorporated psychosocial and humanistic concepts through problem-based learning were more likely to pursue primary care and rated themselves higher on a scale rating their preparation to practice humanistic medicine and ability

[10] See https://www.artsandmindlab.org/. (Accessed October 2, 2017.)

to manage patients with psychosocial problems (Ousager and Johannessen, 2010).

Medicine has been integrated into the arts and humanities in various ways, including through medical illustration, arts therapy, and the field of neuroaesthetics, and as the subject of artistic scholarship and practice in a range of media.

It is also worth noting that the committee observed that the greatest number of available research studies and evaluations on integration reviewed in Chapters 6 fell into two categories: the integration of the arts and humanities into medical training and the integration of the arts and humanities with engineering (see Chapter 6). The availability of these studies may relate to the fact that both medicine and engineering are applied fields that have very direct impacts on people—in contrast, for example, to basic scientific research. Perhaps this direct connection to people and society places a greater demand on these fields to prepare students to understand the humanistic and artistic dimensions of medical and engineering practice.

8

Findings and Recommendations

Curricular approaches in American higher education today have arisen out of a historical tradition of educational integration, originally called liberal education, through which students were exposed to the full suite of human knowledge with the goal of preparing them for work, life, and civic participation. This holistic approach to education offered students a breadth of exposure to the diverse forms of knowledge and inquiry pursued and produced by the different academic disciplines—the humanities, the arts, the sciences, engineering, mathematics, and medicine. It also helped them to understand how this knowledge is connected. The goal was to impress upon students that all forms of human knowledge and inquiry are branches from the same tree.

In the twentieth century, this liberal approach to education evolved. Today students at most schools are still exposed to a broad array of disciplines through general education programs. That said, as the academic disciplines have specialized, and higher education institutions have developed administrative structures that are fragmented along disciplinary lines, some faculty and leaders in higher education are now questioning whether the education we are offering students today allows them to appreciate the connections between the disciplines. Many are now calling for a return to a more integrative approach to education. Proponents of a turn toward a more integrative approach in higher education argue that an education shaped by disciplinary specialization may not best serve the learning and career goals of most students or prepare future generations to address the complex, and often unpredictable, challenges and opportunities that will face the nation and the world in the twenty-first century. Indeed, many institutions of higher education have embraced this idea and have implemented a range of different integrative courses and programs that aim to intentionally connect content and pedagogies across the humanities, arts, natural and physical sciences, social sciences, engineering, technology, mathematics, and medicine (see "Compendium of Programs and Courses That Integrate the Humanities, Arts, and STEMM" available at https://www.nap.edu/catalog/24988 under the Resources tab).

The charge to this committee was to examine the evidence of the impact of educational experiences that integrate the humanities and arts with the sciences, technology, engineering, mathematics, and medicine on both undergraduate and graduate students in terms of learning and career outcomes. To carry out our task we considered a diverse range of integrative programs and courses, some that are relatively new and others that are established at universities and have been running successfully for several decades. Our review of the existing evidence revealed a dearth of causal evidence on the impact of integrative courses and programs on students, which was unsurprising given the challenges of carrying out randomized, controlled, longitudinal research in higher education. But the committee does not consider causal studies the only legitimate form of evidence. We see value in multiple forms of evidence (e.g., narrative, anecdotal, case study, expert opinion, correlational, quasi-experimental, etc.) and acknowledge that the collection of evidence in the real world rarely, if ever, begins with a longitudinal, controlled trial. Rather, evidence is collected in stages and usually begins with observation and description.

After considering multiple forms of evidence, the committee found that certain approaches to the integration of the arts and humanities with science, technology, engineering, mathematics, and medicine (STEMM) are associated with positive student learning outcomes, including, but not limited to, written and oral communication skills, teamwork skills, ethical decision making, critical thinking and deeper learning, content mastery, general engagement and enjoyment of learning, empathy, resilience, the ability to apply knowledge in real-world settings, and indicators of improved science literacy (Ebert-May et al., 2010; Gurnon et al., 2013; Ifenthaler et al., 2015; Jarvinen and Jarvinen, 2012; Krupczak, 2004; Krupczak and Ollis, 2006; Malavé and Watson, 2000; Naghshineh et al., 2008; Olds and Miller, 2004; Ousager and Johannessen, 2010; Pollack and Korol, 2013; Sands et al., 2008; Stolk and Martello 2015; Thigpen et al., 2004; Willson et al., 1995). These learning outcomes are associated with specific studies that have looked at diverse forms of integration (e.g., integration of engineering and history, integration of medicine and art observation, integration of neuroscience and poetry, etc.) that have adopted different pedagogical approaches (e.g., project-based learning, lecture, living-learning community, etc.) and have appeared in the curriculum in different ways (e.g., as a stand-alone course, co-curricular activity, fully integrated program). Given this diversity it is not possible to make generalizations about the impact of integration as a general approach; however, the committee was struck by the fact that many of these learning outcomes are those that higher education institutions and employers agree will prepare students to enter the workforce, help them live enriched lives, and enable them to become active and informed members of a modern democracy.

Given that the available evidence is promising and indicates positive outcomes for students, the committee is urging a new nationwide effort to develop and fund the research agenda needed to collect the robust and multifaceted evidence that the broader educational community can accept, embrace, and apply to specific settings throughout the huge and complex landscape of American higher education. Though more research on the impact of integrative courses and programs is needed, the committee does not believe it is practical for institutions of higher education to wait to support and adopt integrative models. Rather, we recommend that institutions that view an integrative approach as potentially beneficial for their students move forward with the adoption of integrative courses and programs *and evaluate them.*

To be clear, the purpose of this report has not been to critique or reject existing disciplinary structures; the committee agrees that the disciplines remain essential, exceptionally valuable, and generative features of contemporary higher education. Rather, our purpose has been to highlight the extraordinary reservoir of potential that the disciplines represent and to evaluate strategies for harnessing that potential through integration.

The committee's conclusions do point to the need for deep and sustained reflection about how institutional modes of organizing fields of knowledge may constrain or distort the potential for integration, in education and beyond. Furthermore, the committee has recognized that some of the most significant challenges to integration are institutional rather than intellectual. The gravitational pull of the disciplines shapes curricula, pedagogical approaches, scholarly practices, and allocations of institutional resources. At the same time, modest institutional commitments to create space for integrative approaches and scholars have the potential for far-reaching effects. In this regard, integration's grand challenges, like societies, are also its most promising opportunities.

FINDINGS

Chapter 2: Higher Education and the Demands of the Twenty-First Century

1. The tradition of liberal education in the United States is holistic. Despite the diversity of institutional types and the diversity of

approaches and priorities within each institution, many institutions of higher education agree on essential learning outcomes that cut across general education and the majors.

2. There are internal and external pressures that drive the intense disciplinary structure embedded in many institutions and serve as barriers to integration, including the influence of accrediting bodies; practices related to the training, promotion, and tenure of faculty; and budgetary structures.
3. Higher education should prepare graduates for employment that does not narrowly align with the focus of their college major. Graduates should expect to have several jobs—and possibly several careers—in their lifetime.
4. Surveys show that employers value graduates who have both technical depth in a given discipline and cross-cutting "twenty-first century" skills and knowledge, such as critical thinking, communications skills, the ability to work well in teams, ethical reasoning, and creativity.
5. Many employers do not believe that higher education is appropriately preparing graduates for the workplace, while many higher education administrators and faculty do.
6. Surveys show that students' goals are aligned with employers' goals and that students place value on doing well in college and finding a "good job" after college.
7. Increasing enrollment in certain interdisciplinary courses and majors suggests that students are interested in integration.
8. Integration has the potential to better equip students with the knowledge, skills, and competencies required of professionals and citizens in today's complex, multidimensional, and challenging world.

Chapter 3: What Is Integration?

1. The concept of integration is a broad one, encompassing different educational approaches in differing settings. Because of that breadth and variety, a singular, universally applicable definition of integration is difficult to achieve.
2. In the research literature, the term "integration" may refer to educational experiences that bring together content and pedagogies in the arts, humanities, sciences, engineering, and medicine, or it may refer to courses and programs that integrate across other dimensions of learning, such as school and life (e.g., living-learning communities). In this study, we use the term "integration" to refer to the former.

3. The theory and practice of curricular integration and integrative learning are evolving. As such, the evidence base and research methodologies are also developing and changing.
4. Different approaches to integration can lead to different levels of integration, such as interdisciplinary, multidisciplinary, or transdisciplinary integration.
5. Many different types of institutions are embracing new, more integrative approaches.
6. Integrative educational experiences may take place in individual courses (in-course), within integrated curricular programs (within-curriculum), or outside of the formal curriculum (co-curricular).

Chapter 4: The Challenges of Assessing the Impact of Integration in Higher Education on Students

1. Understanding the impact of integrative educational approaches is challenging for several reasons, including the following:
   a. Each discipline approaches evidence and value differently when evaluating the impact of integrative courses and programs.
   b. The wide variety of integrative programs, courses, and approaches may result in a wide variety of learning outcomes.
   c. Different faculty members and institutions may prioritize and evaluate different learning outcomes, even for similar courses or programs.
   d. Faculty or institutions may not have the time, resources, and expertise to evaluate learning outcomes.
   e. Given the newness of many integrative programs and courses, we do not currently have longitudinal data on the impact of such educational experiences on students.
2. We conclude that it is appropriate and necessary to consider multiple forms of evidence when evaluating the impact of an educational experience on a student, and that approaches to evaluating the impact of courses and programs that integrate the humanities, arts, and STEMM will necessarily be diverse and should be aligned with the specific learning goals of the course or program in its own institutional context.
3. Studying the effects of integration in higher education, like studying college student learning more generally, is a highly complex task that challenges precision and determinations of causation. For example, randomized, controlled, experimental approaches are rarely feasible, and even if researchers can isolate an aspect of an integrative experience appropriate for experimental study, they run

the risk of distilling out what was creating the experience's effect in the first place.

4. Higher education research has developed reliable, verifiable tools that provide valuable evidence on the impact of educational experiences on students, and these should continue to be used to elucidate the impact of integrative educational models.

Chapter 5: Understanding and Overcoming the Barriers to Integration in Higher Education

1. Despite enthusiasm for interdisciplinary approaches in teaching and research, numerous challenges tend to discourage interdisciplinary integration—even within related fields. Rigid professional identities, disciplinary structures, and organizational and bureaucratic arrangements are interlinked in ways that tend to disincentivize interdisciplinary integration.
2. Reappointment, promotion, and tenure often require faculty to meet teaching, research, and service expectations, all of which tend to be based on disciplinarily defined criteria and determined by senior members of their discipline. Thus, even though early career faculty are more inclined to be more innovative in research or teaching, they are also vulnerable to the potentially negative professional consequences of doing so.
3. Many faculty members lack the preparation and expertise needed to offer integrative courses. There are, however, some faculty that have been trained in interdisciplinary fields, such as Arts, Media, and Engineering, or Science and Technology Studies, who are well positioned to teach integrative courses.
4. Justifying interdisciplinary courses in budgetary allocations or incorporating them into curricula where accreditation standards disallow or appear to discourage departures from traditional, disciplinary practices may prove to be challenging for departments interested in fostering interdisciplinary integration.
5. Some institutions have begun to change policies to explicitly include interdisciplinary scholarship within the criteria for promotion and tenure. The strategies at Arizona State University, University of Arizona, Rochester Institute of Technology, University of Michigan, and Indiana University Bloomington offer a few examples of how specific institutions have worked to break down the cultural and administrative barriers to integration.

Chapter 6: The Effects of Integration on Students at the Undergraduate Level

1. The aggregate evidence from various sources, including the peer-reviewed literature, suggests that integration of the arts and humanities with STEM at the undergraduate level leads to certain positive learning outcomes, such as critical thinking, communications skills, the ability to work well in teams, content mastery, improved visuospatial skills, and improved motivation and enjoyment of learning. Additional positive outcomes include improved retention, better GPAs, and higher graduation rates.
2. Many of the learning outcomes associated with the integration of the humanities and arts with STEM align with those that employers say they are looking for in recent graduates. These same learning outcomes are valued by many institutions of higher education and should serve students well in their personal and civic lives.
3. Though evidence of the value of integrating STEM into the humanities and arts is limited, a survey of existing courses and programs suggests that such integration can offer students and scholars new tools and frameworks for humanistic and artistic inquiry and can lead to improved scientific and technological literacy among students.
4. A catalog of known integrative programs, although not exhaustive, demonstrates that many integrative programs and courses are taking place at many different types of educational institutions.
5. Emerging evidence suggests that integration positively affects the recruitment, learning, and retention of women and underrepresented minorities in science and engineering.

Chapter 7: Integration in Graduate and Medical Education

1. Federal agencies (e.g., the National Science Foundation, National Institutes of Health, National Endowment for the Humanities, and National Endowment for the Arts) and private funding agencies (e.g., the Andrew W. Mellon and the Alfred P. Sloan Foundations) have offered limited support for programs that integrate STEM fields, the humanities, and the arts. The National Science Foundation's Integrative Graduate Education and Research Traineeship (IGERT) is an example of how funding agencies can promote integrative graduate education.
2. Graduate programs described in the published literature that integrate the humanities, arts, and STEM disciplines report positive

learning outcomes, though much of the evidence for these outcomes is anecdotal.

3. Graduate programs are an important source of exploratory arts–humanities–STEMM integration and have at times led to the creation of new interdisciplinary fields and areas of study (including Bioethics; Science, Technology, and Society; Media Arts; Human–Computer Interaction; and Digital Humanities). These interdisciplinary fields explore questions not addressed within traditional disciplines and tend to focus on issues of social relevance.
4. Interdisciplinary graduate programs can make a valuable contribution to the preparation of future faculty and the development of institutional capacity for delivering integrative undergraduate experiences.
5. In addition to preparing future faculty, graduate training programs are an important element in the larger ecosystem of integrative interdisciplinary teaching and research. We observed that institutions with significant undergraduate integrative education also demonstrated institutional support for graduate training programs and faculty hiring and research support in integrative interdisciplinary fields.
6. Administrative and budgetary structures of higher education may either support or constrain the development of interdisciplinary courses and programs.
7. Crossing the boundaries of disciplines, courses, and programs requires a conscious effort by administrators to provide incentives and support for integration.
8. Scholars in interdisciplinary fields learn how to move between disciplinary approaches, recognize their critical limitations, and integrate them. As such, they represent an important reservoir of experience and expertise about how to pursue integration more broadly.
9. Courses and programs that integrate the arts and humanities with medical and premedical training are associated with improved visual diagnostic skills, resiliency, empathy, and self-efficacy.

Based on our findings, this committee has concluded that higher education should strive to offer all students—regardless of degree or area of concentration—an education that exposes them to diverse forms of human knowledge and inquiry and that impresses upon them that all disciplines are "branches of the same tree." Such an education should empower students to understand the fundamental connections among the diverse branches of human inquiry—the arts, humanities, sciences, social sciences, mathematics, engineering, technology, and medicine.

Though the evidence on the impact of integrative educational approaches is limited, it suggests potential benefits to students that warrant future research. The outcomes associated with various approaches to integration—improved written and oral communications skills, teamwork skills, ethical decision making, critical thinking and deeper learning, content mastery, general engagement and enjoyment of learning, empathy, resilience, the ability to apply knowledge in real-world settings, and indicators of improved science literacy—are encouraging. It is our consensus opinion that integrative approaches in higher education have the potential to benefit graduates in work, life, and civic engagement.

While it is true that the current evidence base limits our ability to draw causal links between integrative curricula in higher education and student learning and career outcomes, given how difficult and time consuming it is to carry out controlled, longitudinal studies in higher education, we do not believe it is practical for institutions with an interest in pursuing more integrative approaches to wait for more robust causal evidence before adopting, supporting, and evaluating integrative programs. This committee has concluded that the available evidence is sufficient to urge support for courses and programs that integrate the arts and humanities with STEMM in higher education. Below we enumerate the specific recommendations that fulfill this vision.

RECOMMENDATIONS

Recommendation 1: This committee has concluded that the available evidence is sufficient to urge the support and evaluation of courses and programs that integrate the arts and humanities with the natural sciences, social sciences, technology, engineering, mathematics, and medicine in higher education. Therefore, we recommend the following:

a. Individual campus departments and schools, campus-wide teams, and campus-employer collaborators should consider developing and implementing new models and programs of STEMM–arts–humanities integration.
b. Developers of integrative programs and courses should include a strong evaluation component to measure the effectiveness of integrative educational models on student learning and workforce readiness. This evaluation component should include, but not be limited to, longitudinal studies of student outcomes and should be developed in collaboration with scholars of higher education research.
c. Funders—including federal agencies, states, and private foundations—should support evaluation of such models and programs.

Recommendation 2: Students should insist on, and institutions should provide, an academic experience that prepares them for life, work, and citizenship in the twenty-first century by strengthening their critical thinking, communications skills, ability to work well in teams, content mastery, motivation, and engagement with learning. Institutions should continue to evaluate and explore the connection between such learning outcomes and integrative curricular models.

Recommendation 3: When working to implement integrative curricular models, institutions should set aside resources for the hiring, research, teaching activities, and professional development of faculty who are capable of teaching integrative courses or programs.

Recommendation 4: Institutions and employers should collaborate to better understand how graduates who participated in courses and programs that integrate the humanities and arts with science, technology, engineering, mathematics, and medicine fare in the workplace throughout their career. Specifically:

a. Institutions should survey alumni to gain a sense of how their education, particularly the integrative aspects of their programs, has served them in work, life, and civic engagement. Institutions should share the results of such surveys with employers.
b. Employers should gather and share with institutions information about the educational experiences, especially integrative experiences, that lead to employee success.
c. Where possible, institutions and employers should find ways to collaborate on these activities.

Recommendation 5: Professional artistic, humanistic, scientific, and engineering societies should work together to build, document, and study integrative pilot programs and models to support student learning and innovative scholarship at the intersection of disciplines.

Recommendation 6: Proponents of disciplinary integration in higher education, including faculty and administrators, should work with scholars of higher education and experts in the humanities, arts, natural sciences, social sciences, engineering, technology, mathematics, and medicine to establish agreement on the expected learning outcomes of an integrative educational experience and work to design scalable, integrative approaches to assessment.

Recommendation 7: Stakeholders (e.g., faculty, administrators, and scholars of higher education research) should employ multiple forms

of inquiry and evaluation when assessing courses and programs that integrate the humanities, arts, natural sciences, social sciences, technology, engineering, mathematics, and medical disciplines, including qualitative, quantitative, narrative, expert opinion, and portfolio-based evidence. Stakeholders should also consider developing new evaluation methodologies for integrative courses and programs.

Recommendation 8: Given the challenges of conducting controlled, randomized, longitudinal research on integrative higher education programs, we recommend two potential ways forward: (1) institutions with specific expertise in student learning outcomes (e.g., schools of education) could take a leadership role in future research endeavors, and (2) several institutions could form a multisite collaboration under the auspices of a national organization (e.g., a higher education association) to carry out a coordinated research effort. In either case, efforts to identify the appropriate expertise and support necessary to conduct such research should be a priority.

Recommendation 9: Institutions should perform a cultural audit of courses, programs, and spaces on campus where integration is already taking place, partnering with student affairs professionals to evaluate programs and initiatives intended to integrate learning between classroom and nonclassroom environments, and working with teaching and learning centers to develop curricula for faculty charged with teaching for or within an integrative experience.

Recommendation 10: Further research should focus on how integrative educational models can promote the representation of women and underrepresented minorities in specific areas of the natural sciences, social sciences, technology, engineering, mathematics, medicine, arts, and humanities, and all research efforts should account for whether the benefits of an integrative approach are realized equitably.

Recommendation 11: Institutions should work to sustain ongoing integrative efforts that have shown promise, including but not limited to, new integrative models of general education.

Recommendation 12: New designs for general education should consider incorporating interdisciplinary and transdisciplinary integration, emphasizing applied and engaged learning and connections between general education and specialized learning throughout the undergraduate years and across the arts, humanities, and STEMM disciplines.

Recommendation 13: When implementing integrative curricula, faculty, administrators, and accrediting bodies need to explore, identify, and mitigate constraints (e.g., tenure and promotion criteria, institutional budget models, workloads, accreditation, and funding sources) that hinder integrative efforts in higher education.

Recommendation 14: Academic thought leaders working to facilitate integrative curricular models should initiate conversations with the key accrediting organizations for STEMM, the arts, and higher education to ensure that the disciplinary structures and mandates imposed by the accreditation process do not thwart efforts to move toward more integrative program offerings.

Recommendation 15: Both federal and private funders should recognize the significant role they can and do play in driving integrative teaching, learning, and research. We urge funders to take leadership in supporting integration by prioritizing and dedicating funding for novel, experimental, and expanded efforts to integrate the arts, humanities, and STEMM disciplines. Sustained support will be necessary to realize the long-term impact of new approaches to disciplinary integration.

Recommendation 16: Interdisciplinarity adds an additional layer of complexity to pedagogy. Professional development of current and future faculty is necessary to promote interdisciplinary teaching and learning. Additional research on effective pedagogical practices for interdisciplinary learning is needed.

Epilogue

The Einstein quote that opens this report suggests that a tree and its branches serve as a metaphor for thinking about the integration of arts, engineering, humanities, mathematics, medicine, science, and technology because the vitality of the whole depends on the combined force of the parts. The trunk of the tree represents the core strength of the disciplines in higher education—the centralizing force that directs students through

the course of academic study. Yet the branches—where Einstein located religion, arts, and sciences—could also be seen as the locations for integration, as they move away from the trunk yet remain integrally connected to the core strengths of the whole. Most importantly, the branches create opportunities for trees to connect to each other. In a forest, the canopy of intersecting branches connects distinct units. In this metaphor, it is the connections between branches and trunk (and roots), rather than the singular strength of any one part, that make the tree healthy and viable.

The statue of Einstein on the grounds of the National Academies of Sciences, Engineering, and Medicine faces the National Mall, not far from the Vietnam Veterans Memorial, the Lincoln Memorial, and the Reflecting Pool. Immediately behind the statue is a small grove of trees, which provide some shade in the summer (and backdrops to countless photographs of visitors sitting on Einstein's lap). These trees provide a fitting image for this study of integration, for it is the connections between the flourishing branches, rather than the spindly trunks, that creates the canopy that provides both shade in the summer and viability in poor weather. Thinking about integration as the connections between branches transforms Einstein's metaphor—as well as the setting of his monument—into a meaningful guide for future action.

References

AAC&U (Association of American Colleges and Universities). 2002. Greater Expectations: A New Vision for Learning as a Nation Goes to College. Washington, DC: AAC&U. https://www.aacu.org/sites/default/files/files/publications/GreaterExpectations.pdf.

AAC&U. 2007. College Learning for the New Global Century. Washington, DC: AAC&U. https://www.aacu.org/sites/default/files/files/LEAP/GlobalCentury_final.pdf.

AAC&U. 2010. Integrative and applied learning value rubric. https://www.aacu.org/value/rubrics/integrative-learning (accessed October 9, 2017).

AAMC. 2008. Recommendations for Clinical Skills Curricula for Undergraduate Medical Education. https://members.aamc.org/eweb/upload/Recommendations%20for%20Clinical%20Skills%20Curricula%202005.pdf.

Abbott, A. 2001. Chaos of disciplines. University of Chicago Press.

ABET. 2018. Accreditation Policy and Procedure Manual (APPM) 2018-1019. http://www.abet.org/accreditation/accreditation-criteria/accreditation-policy-and-procedure-manual-appm-2018-2019/.

Ainsworth, S., V. Prain, and R. Tytler. 2011. Drawing to learn in science. *Science 333*(6046), 1096-1097.

Akcay, B., and H. Akcay. 2015. Effectiveness of science-technology-society (STS) instruction on student understanding of the nature of science and attitudes toward science. *International Journal of Education in Mathematics, Science and Technology* 3(1):37-45.

Albertine, S. 2012. Liberal education in community colleges. The LEAP Challenge blog. https://www.aacu.org/leap/liberal-education-nation-blog/liberal-education-community-colleges (accessed October 4, 2017).

Alias, M., T. R. Black, and D. E. Gray. 2002. Effect of instruction on spatial visualization ability in civil engineering students. *International Education Journal* 3(1).

Allen, D. 2004. *Talking to Strangers: Anxieties of Citizenship Since Brown v. Board of Education*: University of Chicago Press.

American Society for Engineering Education. Case studies. https://www.asee.org/engineering-enhanced-liberal-education-project/case-studies (accessed October 3, 2017).

Arntfield, S. L., K. Slesar, J. Dickson, and R. Charon. 2013. Narrative medicine as a means of training medical students toward residency competencies. *Patient Education and Counseling* 91(3):280-286.

Asbury, C. H., and B. Rich. 2008. *Learning, Arts, and the Brain: The Dana Consortium Report on Arts and Cognition*. New York/Washington, DC: Dana Press.

Association of American Colleges. 1994. *Strong Foundations: Twelve Principles for Effective General Education Programs*. Washington, DC: Association of American Colleges.

Barber, J. P. 2012. Integration of learning: A grounded theory analysis of college students' learning. *American Educational Research Journal* 49(3):590-617.

Barr, R. B., and J. Tagg. 1995. From teaching to learning—a new paradigm for undergraduate education. *Change: The Magazine of Higher Learning* 27(6):12-26.

Bass, R., and B. Eynon. 2017. From unbundling to rebundling: Design principles for transforming institutions in the new digital ecosystem. *Change: The Magazine of Higher Learning* 49(2):8-17.

Begg, M. D., L. M. Bennett, L. Cicutto, H. Gadlin, M. Moss, J. Tentler, and E. Schoenbaum. 2015. Graduate education for the future: New models and methods for the clinical and translational workforce. *Clinical and Translational Science* 8(6):787-792.

Begg, M. D., G. Crumley, A. M. Fair, C. A. Martina, W. T. McCormack, C. Merchant, C. M. Patino-Sutton, and J. G. Umans. 2014. Approaches to preparing young scholars for careers in interdisciplinary team science. *Journal of Investigative Medicine* 62(1):14-25.

Begg, M. D., and R. D. Vaughan. 2011. Are biostatistics students prepared to succeed in the era of interdisciplinary science? (and how will we know?). *American Statistician* 65(2):71-79.

Bennett, C. 2014. Anatomic body painting: Where visual art meets science. *The Journal of Physician Assistant Education* 25(4):52-54.

Berry, S., E. G. Lamb, and T. Jones. 2016. Health humanities baccalaureate programs in the United States. *Center for Literature and Medicine, Hiram College*. http://www. hiram. edu/images/pdfs/center-litmed/HHBP_8_11_16. pdf.

Blue, E., M. Levine, and D. Nieusma. 2013. *Engineering and War: Militarism, Ethics, Institutions, Alternatives*. San Rafael, CA: Morgan & Claypool Publishers.

Bodenhamer, D. J., J. Corrigan, and T. M. Harris. 2010. *The Spatial Humanities: GIS and the Future of Humanities Scholarship*. Bloomington: Indiana University Press.

Bohannon, J. 2016. And the winner of this year's dance your Ph.D. contest is.... *Science*. http://www.sciencemag.org/news/2016/10/and-winner-year-s-dance-your-phd-contest (accessed September 28, 2017).

Bonner, T. N. 1963. *American Doctors and German Universities: A Chapter in International Intellectual Relations, 1870-1914*. Lincoln: University of Nebraska Press.

Borrego, M., and S. Cutler. 2010. Constructive alignment of interdisciplinary graduate curriculum in engineering and science: An analysis of successful igert proposals. *Journal of Engineering Education* 99(4):355-369.

Borrego, M., C. Newswander, L. McNair, S. McGinnis, and M. Paretti. 2009. Using concept maps to assess interdisciplinary integration of green engineering knowledge. *Advances in Engineering Education* 1(3).

Borrego, M., and L. K. Newswander. 2010. Definitions of interdisciplinary research: Toward graduate-level interdisciplinary learning outcomes. *Review of Higher Education* 34(1):61-84.

Botstein, L. 2000. The training of musicians. *The Musical Quarterly* 84(3):327-332.

Bowman, N. A. 2010. Disequilibrium and resolution: The nonlinear effects of diversity courses on well-being and orientations toward diversity. *The Review of Higher Education* 33(4):543-568.

Boylston, A. W. 2018. The myth of the milkmaid. *New England Journal of Medicine* 378(5):414-415.

Braverman, I. M. 2011. To see or not to see: how visual training can improve observational skills. *Clinics in Dermatology* 29(3):343-346.

Breithaupt, F. 2017. Designing a lab in the humanities. *The Chronicle of Higher Education.* http://www.chronicle.com/article/Designing-a-Lab-in-the/239132 (accessed October 9, 2017).

Brown, A. S., and S. J. Tepper. 2012. *Placing the Arts at the Heart of the Creative Campus.* New York: Association of Performing Arts Presenters.

Brown, J. S., A. Collins, and P. Duguid. 1989. Situated cognition and the culture of learning. *Educational Researcher* 18(1):32-42.

Brown, S., and J. Giordan. 2008. *IGERT Integrative Graduate Education and Research Traineeship Annual Report 2006–2007.* Arlington, VA: National Science Foundation.

Brown, T. J., and D. F. Kuratko. 2015. The impact of design and innovation on the future of education. *Psychology of Aesthetics, Creativity, and the Arts* 9(2):147-151.

Brunner, C. 1997. Opening technology to girls. *Electronic Learning* 16(4):55.

Bucher, K. 2016. New frontiers of medical illustration. *JAMA* 316(22):2340-2341.

Bullough, R. V., Jr. 2006. Developing interdisciplinary researchers: What ever happened to the humanities in education? *Educational Researcher* 35(8):3-10.

Bureau of Labor Statistics. 2016. Employee Tenure in 2016. https://www.bls.gov/news.release/pdf/tenure.pdf.

Burning Glass Technologies. 2015. The Human Factor: the Hard Time Employers Have Finding Soft Skills. Boston, MA: Burning Glass Technologies. https://www.burning-glass.com/wp-content/uploads/Human_Factor_Baseline_Skills_FINAL.pdf.

Busteed, B. 2014. Higher education's work preparation paradox. *Gallup.* http://news.gallup.com/opinion/gallup/173249/higher-education-work-preparation-paradox.aspx.

Byars-Winston, A., Y. Estrada, C. Howard, D. Davis, and J. Zalapa. (2010). Influence of social cognitive and ethnic variables on academic goals of underrepresented students in science and engineering: a multiple-groups analysis. *Journal of Counseling Psychology* 57(2):205.

Calatrava Moreno, M. D. C., and M. A. Danowitz. 2016. Becoming an interdisciplinary scientist: An analysis of students' experiences in three computer science doctoral programmes. *Journal of Higher Education Policy and Management* 38(4):448-464.

Calderon, V. J., and P. Sidhu. 2014. Business leaders say knowledge trumps college pedigree. *Gallup.* http://news.gallup.com/poll/167546/business-leaders-say-knowledge-trumps-college-pedigree.aspx.

Calhoun, C., and D. R. Rhoten. 2010. Integrating the social sciences: Theoretical knowledge, methodological tools, and practical applications. Pp. 103-118 in *The Oxford Handbook of Interdisciplinarity*, R. Frodeman, R. Carlos Dos Santos Pacheco, and J. Thompson Klein, eds. Oxford, UK: Oxford University Press.

Carnevale, A. P., N. Smith, and M. Melton. 2011. STEM: Science Technology Engineering Mathematics. Washington, DC: Georgetown University Center on Education and the Workforce.

Carney, J., A. Martinez, J. Dreier, K. Neishi, and A. Parsad (Abt Associates). 2011. Evaluation of the national science foundation's integrative graduate education and research traineeship program (IGERT): Follow-up study of IGERT graduates. Final report. http://www.abtassociates.com/reports/ES_IGERT_FOLLOWUP%20STUDY_FULLREPORT_May_2011.pdf.

Catterall, J. S. 2012. The Arts and Achievement in At-Risk Youth: Findings from Four Longitudinal Studies. Research report #55. Washington, DC: National Endowment for the Arts.

Cavanagh, S. T. 2010. Bringing Our Brains to the Humanities Increasing the Value of Our Classes While Supporting Our Futures. *Pedagogy* 10(1):131-142.

Chamany, K. 2006. Science and social justice. *Journal of College Science Teaching* 36(2):54.

Chaves, C. A. U. 2014. *Liberal Arts and Sciences: Thinking Critically, Creatively, and Ethically.* Bloomington, IN: Trafford Publishing.

Choi, B. C., and A. W. Pak. 2006. Multidisciplinarity, interdisciplinarity and transdisciplinarity in health research, services, education and policy: 1. Definitions, objectives, and evidence of effectiveness. *Clinical and Investigative Medicine* 29(6):351.

Choy, S. P. (2002). Access and Persistence: Findings from 10 Years of Longitudinal Research on Students. Washington, DC: American Council on Education. https://www.michigan.gov/documents/mistudentaid/2002AccessAndPersistence_394481_7.pdf.

Chuk, E., R. Hoetzlein, D. Kim, and J. Panko. 2012. Creating socially networked knowledge through interdisciplinary collaboration. *Arts and Humanities in Higher Education* 11(1-2):93-108.

Conant, J. B. 1950. *General Education in a Free Society, Report of the Harvard Committee.* Cambridge, MA: Harvard University Press.

Confino, J., and L. Paddison. 2014. Cookstove Designs are Failing the Poorest Communities. London, UK: Guardian News & Media Ltd. Available online at: http://www.theguardian.com/sustainable-business/cookstoves-design-poorcommunities-refugees-unhcr-ikea (accessed September 12, 2015).

Crawford, M. B. 2009. *Shop Class as Soulcraft: An Inquiry into the Value of Work.* New York: Penguin Press.

Crombie, A. C. 1994. Styles of scientific thinking in the European tradition: The history of argument and explanation especially in the mathematical and biomedical sciences and arts (Vol. 2). Duckworth.

Crow, M. M., and W. B. Dabars. 2014. Interdisciplinarity as a design problem: Toward mutual intelligibility among academic disciplines in the American research university. Enhancing Communication and Collaboration in Interdisciplinary Research, Sage Publications, Los Angeles, pp. 294-322.

Crowther, G. 2012. Using science songs to enhance learning: An interdisciplinary approach. *CBE Life Sciences Education* 11(1):26-30.

Curşeu, P. L., and H. Pluut. 2013. Student groups as learning entities: The effect of group diversity and teamwork quality on groups' cognitive complexity. *Studies in Higher Education* 38(1):87-103.

Dail, W. 2013. On cultural polymathy: How visual thinking, culture, and community create a platform for progress. *The STEAM Journal* 1(1):7.

Dalbello, M. 2011. A genealogy of digital humanities. *Journal of Documentation* 67(3):480-506.

David, D. 2015. Giving California students a compass. https://www.aacu.org/sites/default/files/files/publications/GivingCaliforniaStudentsaCompass.pdf (accessed October 4, 2017).

Deegan, M. 2014. This ever more amorphous thing called digital humanities': Whither the humanities project? *Arts and Humanities in Higher Education* 13(1-2):24-41.

Delaney, J. A., and S. Suissa. 2009. The case-crossover study design in pharmacoepidemiology. *Statistical Methods in Medical Research* 18(1):53-65.

Delbanco, A. 2012. *College: What It Was, Is, and Should Be.* Princeton, NJ: Princeton University Press.

Denbo, S. M. 2008. Nature's business: Incorporating global studies, environmental law and literacy, and corporate social responsibility into the business school curriculum through interdisciplinary "business-science" study tour courses. *Journal of Legal Studies Education* 25(2):215-240.

Deno, J. A. 1995. The relationship of previous experiences to spatial visualization ability. *Engineering Design Graphics Journal* 59(3):5-17.

DiBiasio, D., P. Quinn, K. Boudreau, L. A. Robinson, J. M. Sullivan Jr., J. Bergendahl, and L. Dodson. "The Theatre of Humanitarian Engineering." In ASEE Annual Conference. 2017.

Dick, R. W., J. R. Berning, W. Dawson, R. D. Ginsburg, C. Miller, and G. T. Shybut. 2013. Athletes and the arts—the role of sports medicine in the performing arts. *Current Sports Medicine Reports* 12(6):397-403.

Diehl, C. 1978. *Americans and German Scholarship, 1770-1870*: New Haven, CT: Yale University Press.

DiLalla, L. F., S. K. Hull, and J. K. Dorsey. 2004. Effect of gender, age, and relevant course work on attitudes toward empathy, patient spirituality, and physician wellness. *Teaching and Learning in Medicine* 16(2):165-170.

Dolev, J. C., L. K. Friedlaender, and I. M. Braverman. 2001. Use of fine art to enhance visual diagnostic skills. *JAMA* 286(9):1020-1021.

Ebert-May, D., E. Bray Speth, J. L. Momsen, J. Hildebrand, and J. Meinwald. 2010. Assessing scientific reasoning in a liberal learning curriculum. *Science and the Educated American: A Core Component of Liberal Education* 228.

Economist, T. 2017. *Climate change and inequality: The rich pollute, the poor suffer*. https://www.economist.com/news/finance-and-economics/21725009-rich-pollute-poor-suffer-climate-change-and-inequality (accessed October 10, 2017).

Einstein, A. 1956. Out of my later years. Citadel Press.

Einstein, A. 2006. *The Einstein Reader*. New York: Citadel Press.

Einstein, A. 2015. *Out of My Later Years: The Scientist, Philosopher, and Man Portrayed Through His Own Words*. New York: Open Road.

Engerman, D. C. 2015. The pedagogical purposes of interdisciplinary social science: A view from area studies in the united states. *Journal of the History of the Behavioral Sciences* 51(1):78-92.

Ernest, J. B., and R. Nemirovsky. 2015. Arguments for integrating the arts: Artistic engagement in an undergraduate foundations of geometry course. *PRIMUS* 26(4):356-370.

Everett, L. J., P. K. Imbrie, and J. Morgan. 2000. Integrated curricula: Purpose and design. *Journal of Engineering Education* 89(2):167-175.

Facione, P. A., and N. C. Facione. 1992. *The California Critical Thinking Disposition Inventory (CCTDI): A Test of Critical Thinking Disposition*. Millbrae: California Academic Press.

Fantauzzacoffin, J., J. D. Rogers, J. D. Bolter, and I.n.I.S.E. Conference. 2012. From STEAM research to education: An integrated art and engineering course at georgia tech.1-4.

Feller, I. 2002. New organizations, old cultures: strategy and implementation of interdisciplinary programs. *Research Evaluation* 11(2):109-116.

Fischer, K. 2012. In Rhode Island, an unusual marriage of engineering and languages lures students. *Chronicle of Higher Education*. https://www.chronicle.com/article/an-unusual-marriage-of/131905.

Fisher, J. A. 2011. *Gender and the Science of Difference: Cultural Politics of Contemporary Science and Medicine*. Newark, NJ: Rutgers University Press.

Fleck, L. 1981. Genesis and Development of a Scientific Fact, Chicago: University of Chicago Press.

Fraser, D. W. 1987. Epidemiology as a liberal art. *New England Journal of Medicine* 316(6):309-314.

Gaff, J. G. 1991. *New Life for the College Curriculum: Assessing Achievements and Furthering Progress in the Reform of General Education*. San Francisco, CA: Jossey-Bass.

Gallup and Purdue University. 2014. Great Jobs Great Lives: The 2014 Gallup-Purdue Index Report. http://news.gallup.com/reports/197141/4.aspx (accessed October 9, 2017).

Gallup and Purdue University. 2015. Gallup-Purdue Index Report 2015. http://news.gallup.com/reports/197144/gallup-purdue-index-report-2015.aspx (accessed October 10, 2017).

Ge, X., D. Ifenthaler, and J. M. Spector. 2015. Moving forward with STEAM education research. Pp. 383-395 in *Emerging Technologies for STEAM Education: Full Steam Ahead,* X. Ge, D. Ifenthaler, and J. Spector, eds. Cham, Switzerland: Springer.

Gerrans, N. P., and N. K. Hayles. 1999. How we became posthuman—virtual bodies in cybernetics, literature and informatics. *TLS, the Times Literary Supplement* (5023):32.

Ghanbari, S. 2014. Integration of the Arts in STEM: A Collective Case Study of Two Interdisciplinary University Programs. Ed.D. dissertation. University of California, San Diego and California State University, San Marcos. https://escholarship.org/uc/item/9wp9x8sj.

Ghanbari, S. 2015. Learning across disciplines: A collective case study of two university programs that integrate the arts with STEM. *International Journal of Education & the Arts* 16(7).

Ginsberg, A. E. 2008. Triumphs, Controversies, and Change: Women's Studies 1970s to the Twenty-First Century. In The Evolution of American Women's Studies (pp. 9-37). Palgrave Macmillan, New York.

Glynn, S. M. and D. Muth. 1994. Reading and writing to learn science: Achieving scientific literacy. *Journal of Research in Science Teaching* 31:1057-1073.

Gonzalo, J. D., E. Baxley, J. Borkan, M. Dekhtyar, R. Hawkins, L. Lawson, S. R. Starr, and S. Skochelak. 2016. Priority areas and potential solutions for successful integration and sustainment of health systems science in undergraduate medical education. *Academic Medicine* 92(1):63-69.

Gonzalo, J. D., M. Dekhtyar, S. R. Starr, J. Borkan, P. Brunett, T. Fancher, J. Green, S. J. Grethlein, C. Lai, L. Lawson, S. Monrad, P. O'Sullivan, M. D. Schwartz, and S. Skochelak. 2017. Health systems science curricula in undergraduate medical education: Identifying and defining a potential curricular framework. *Academic Medicine* 92(1):123-131.

Graff, N. 2011. "An effective and agonizing way to learn": Backwards design and new teachers' preparation for planning curriculum. *Teacher Education Quarterly* 38(3):151-168.

Graff, H. J. 2015. *Undisciplining Knowledge: Interdisciplinarity in the Twentieth Century.* Baltimore, MD: JHU Press.

Grant, J., and D. Patterson. 2016. Innovative arts programs require innovative partnerships: A case study of steam partnering between an art gallery and a natural history museum. *The Clearing House* 89(4-5):144-152.

Grasso, D., and D. Martinelli. 2010. Pp. 81-92 in *Holistic Engineering Education: Beyond Technology,* D. Grasso and M. Burkins, eds. New York: Springer Verlag.

Graybill, J. K., S. Dooling, V. Shandas, J. Withey, A. Greve, and G. L. Simon. 2006. A rough guide to interdisciplinarity: Graduate student perspectives. *BioScience* 56(9):757-763.

Green, M. J., and K. R. Myers. 2010. Graphic medicine: Use of comics in medical education and patient care. *The Veterinary Record: Journal of the British Veterinary Association.* 166(11):574-577.

Groenendijk, T., T. Janssen, G. Rijlaarsdam, and H. van den Bergh. 2013. Learning to be creative. The effects of observational learning on students' design products and processes. *Learning and Instruction* 28:35-47.

Guay, F., R. J. Vallerand, and C. Blanchard. 2000. On the assessment of situational intrinsic and extrinsic motivation: The situational motivation scale (SIMS). *Motivation and Emotion* 24:175-214.

Gülbahar, Y., and H. Tinmaz. 2006. Implementing project-based learning and e-portfolio assessment in an undergraduate course. *Journal of Research on Technology in Education* 38(3):309-327.

Gulbrandsen, M., and S. Aanstad. 2015. Is innovation a useful concept for arts and humanities research? *Arts and Humanities in Higher Education* 14(1):9-24.

Gurnon D., J. Voss-Andreae, J. Stanley. 2013. Integrating art and science in undergraduate education. *PLOS Biology* 11(2): e1001491.

Hackett, E. J., and D. R. Rhoten. 2009. The snowbird charrette: Integrative interdisciplinary collaboration in environmental research design. *Minerva: A Review of Science, Learning and Policy* 47(4):407-440.

Hacking, I. 1992. 'Style' for historians and philosophers. Studies in History and Philosophy of Science Part A, 23(1):1-20.

Hall, K. L., A. L. Vogel, B. A. Stipelman, D. Stokols, G. Morgan, and S. Gehlert. 2012. A four-phase model of transdisciplinary team-based research: Goals, team processes, and strategies. *Translational Behavioral Medicine* 2(4):415-430.

Han, H., and C. Jeong. 2014. Improving epistemological beliefs and moral judgment through an STS-based science ethics education program. *Science and Engineering Ethics* 20(1):197-220.

Haraway, D. J. 1994. A game of cat's cradle: Science studies, feminist theory, cultural studies. *Configurations* 2(1):59.

Harper, J. D., S. Kaiser, K. Ebrahimi, G. R. Lamberton, H. R. Hadley, H. C. Ruckle, and D. D. Baldwin, 2007. Prior video game exposure does not enhance robotic surgical performance. *Journal of Endourology* 21(10):1207-1210.

Hart Research Associates. 2013. It takes more than a major: employer priorities for college learning and student success. Liberal Education, 99.

Hart Research Associates. 2016. Recent Trends in General Education Design, Learning Outcomes, and Teaching Approaches. Key Findings from a Survey Among Administrators at AAC&U Member Institutions. https://www.aacu.org/sites/default/files/files/LEAP/2015_Survey_Report2_GEtrends.pdf.

Hestenes, D., and M. Wells. 1992. A mechanics baseline test. *Physics Teacher* 30(3):159-166.

Hestenes, D., M. Wells, and G. Swackhamer. 1992. Force concept inventory. *Physics Teacher* 30(3):141-158.

Hirt, J. B. 2006. *Where You Work Matters: Student Affairs Administration at Different Types of Institutions*. Lanham, MD: University Press of America.

Ho, P., J. C. I. Tsao, L. Bloch, and L. K. Zeltzer. 2011. The impact of group drumming on social-emotional behavior in low-income children. *Evidence-Based Complementary and Alternative Medicine* 2011(2):1-14.

Hong, L., and S. E. Page. 2004. Groups of diverse problem solvers can outperform groups of high-ability problem solvers. *Proceedings of the National Academy of Sciences* 101(46):16385-16389. http://www.pnas.org/content/101/46/16385.short (accessed October 10, 2017).

Hothem, T. 2013. Integrated General Education and the Extent of Interdisciplinarity: The University of California–Merced's Core 1 Curriculum. *The Journal of General Education* 62(2-3):84-111.

Hull, T. 2006. *Project Origami: Activities for Exploring Mathematics*. Wellesley, MA: A.K. Peters.

IBM. 2009. IBM's Role in Creating the Workforce of the Future. http://www-05.ibm.com/de/ibm/engagement/university_relations/pdf/Beyond_IT_report_IBM_Workforce_of_the_Future.pdf (accessed October 10, 2017).

Ifenthaler, D., Z. Siddique, and F. Mistree. 2015. Designing for open innovation: Change of attitudes, self-concept, and team dynamics in engineering education. Pp. 201-215 in *Emerging Technologies for STEM Education: Full Steam Ahead*, X. Ge, D. Ifenthaler, and J. M. Spector, eds. Cham, Switzerland: Springer.

Inkelas, K. K., K. E. Vogt, S. D. Longerbeam, J. Owen, and D. Johnson. 2006. Measuring outcomes of living-learning programs: Examining college environments and student learning and development. *The Journal of General Education* 55(1):40-76.

Institute of Medicine. 2003. *Health Professions Education: A Bridge to Quality*. Washington, DC: The National Academies Press. https://doi.org/10.17226/10681.

Islam, S. N., and J. Winkel. 2016. UN/DESA Policy Brief #45: The Nexus Between Climate Change and Inequalities. https://www.un.org/development/desa/dpad/wp-content/uploads/sites/45/publication/WESS2016-PB2.pdf (accessed October 10, 2017).

Jarvinen, M. K., and L. Z. Jarvinen. 2012. Elevating student potential: Creating digital video to teach neurotransmission. *Journal of Undergraduate Neuroscience Education* 11(1):A6-A7.

Jasanoff, S. 2010. A field of its own: The emergence of science and technology studies. Pp. 103-118 in *The Oxford Handbook of Interdisciplinarity*, R. Frodeman, R. Carlos Dos Santos Pacheco, and J. Thompson Klein, eds. Oxford, UK: Oxford University Press.

Jaschke, A. C., L. H. Eggermont, H. Honing, and E. J. Scherder. 2013. Music education and its effect on intellectual abilities in children: a systematic review. *Reviews in the Neurosciences* 24(6):665-675. doi: 10.1515/revneuro-2013-0023.

Jeffers, C. S. 2009. Within connections: Empathy, mirror neurons, and art education—preservice art educators explore the question: "Are our brains hardwired for empathic connection and aesthetic response?" *Art Education* 62(2):18.

Jones, C. 2009. Interdisciplinary Approach—Advantages, Disadvantages, and the Future Benefits of Interdisciplinary Studies. http://dc.cod.edu/cgi/viewcontent.cgi?article=1121&context=essai (accessed October 9, 2017).

Kao, C. H.-L. 2017. Duoskin rolls out on the runway. https://www.media.mit.edu/posts/duoskin-rolls-out-on-the-runway/ (accessed October 4, 2017).

Mack, K., M. Soto, L. Casillas-Martinez, and E. F. McCormack. 2015. Women in computing: The imperative of critical pedagogical reform. *AAC&U*. http://aacu.org/diversity democracy/2015/spring/mack (accessed October 3, 2017).

Kerr, C. 2001. *The Uses of the University*. Cambridge, MA: Harvard University Press.

Keup, J. R. 2013. Living-Learning Communities as a High-Impact Educational Practice. http://sc.edu/fye/research/research_presentations/files/Keup_ACUHOI_2013_Providence,RI.pdf (accessed October 10, 2017).

Kirklin, D., J. Duncan, S. McBride, S. Hunt, and M. Griffin. 2007. A cluster design controlled trial of arts-based observational skills training in primary care. *Medical Education* 41(4):395-401.

Klein, J. T. 1990. *Interdisciplinarity: History, Theory, and Practice*. Detroit, MI: Wayne State University Press.

Kline, S. J. 1995. *Conceptual Foundations for Multidisciplinary Tthinking*. Stanford, CA: Stanford University Press.

Klein, J. T. 2009. *Creating Interdisciplinary Campus Cultures: A Model for Strength and Sustainability*. Edison, NJ: John Wiley & Sons.

Klugman, C. M., and D. Beckmann-Mendez. 2015. One thousand words: Evaluating an interdisciplinary art education program. *Journal of Nursing Education* 54(4):220-223.

Klugman, C. M., J. Peel, and D. Beckmann-Mendez. 2011. Art rounds: teaching interprofessional students visual thinking strategies at one school. *Academic Medicine* 86(10):1266-1271.

Krajcik, J., K. L. McNeill, and B. J. Reiser. 2008. Learning-goals-driven design model: Developing curriculum materials that align with national standards and incorporate project based pedagogy. *Science Education* 92(1):1-32.

Krupczak, J. 2004. Reaching out across campus: engineers as champions of technological literacy. WPI Studies, 23:171-188.

Krupczak, J., and D. Ollis. 2006. Technological literacy and engineering for non-engineers: Lessons from successful courses. Paper read at Proceeding of the 2006 American Society for Engineering Education Annual Conference.

Krupczak, J. J., S. VanderStoep, L. Wessman, N. Makowski, C. A. Otto, and K. V. Dyk. 2005. *Work in progress - case study of a technological literacy and non-majors engineering course*.

Kuh, G. D. 2008. *High-Impact Educational Practices: What They Are, Who Has Access to Them, and Why They Matter*. Washington, DC: AAC&U.

Kuhn, T. S. 2012. *The Structure of Scientific Revolutions*. Chicago, IL: University of Chicago Press.

Labov, J. B., A. H. Reid, and K. R. Yamamoto. 2010. From the national academies: Integrated biology and undergraduate science education: A new biology education for the twenty-first century? *CBE-Life Sciences Education* 9(1):10-16.

Laird, T., R. Shoup, and G. D. Kuh. 2005. Measuring deep approaches to learning using the national survey of student engagement. Paper read at annual meeting of the Association for Institutional Research.

LaMore, R., R. Root-Bernstein, M. Root-Bernstein, J. H. Schweitzer, J. L. Lawton, E. Roraback, A. Peruski, M. VanDyke, and L. Fernandez. 2013. Arts and crafts: Critical to economic innovation. *Economic Development Quarterly* 27(3):221-229.

Lampert, N. 2006. Critical thinking dispositions as an outcome of art education. *Studies in Art Education* 47(3):215.

Land, M. H. 2013. Full steam ahead: The benefits of integrating the arts into STEM. *Procedia Computer Science* 20:547-552.

Lang, R. J. 2012. *Origami Design Secrets: Mathematical Methods for an Ancient Art*. Boca Raton: A K Peters/CRC Press.

LCME. 2016. Functions and Structure of a Medical School. http://lcme.org/publications/#Standards.

Leider, J. P., B. C. Castrucci, C. M. Plepys, C. Blakely, E. Burke, and J. B. Sprague. 2015. Characterizing the growth of the undergraduate public health major: U.S., 1992-2012. *Public Health Reports* 130(1):104-113.

Leppa, C. J., and L. M. Terry. 2004. Reflective practice in nursing ethics education: International collaboration. *Journal of Advanced Nursing* 48(2):195-202.

Lewin, L. O., C. A. Olson, K. W. Goodman, and P. K. Kokotailo. 2004. UME-21 and teaching ethics: A step in the right direction. *Family Medicine* 36:S36-S42.

Lewis, A. L. 2015. Putting the h in STEAM: Paradigms for modern liberal arts education. Pp. 259-275 *Emerging Technologies for STEAM Education: Full Steam Ahead*, X. Ge, D. Ifenthaler, and J. Spector, eds. Cham, Switzerland: Springer.

Lipson, A., A. W. Epstein, R. Bras, and K. Hodges. 2007. Students perceptions of terrascope, a project-based freshman learning community. *Journal of Science Education and Technology* 16(4):349-364.

Liu, X., and Y. Qian. 2017. If you were a tree. https://www.media.mit.edu/posts/tree-treesense/ (accessed October 4, 2017).

Lobban, J. 2014. The invisible wound: Veterans' art therapy. *International Journal of Art Therapy* 19(1):3-18.

Lucas, C. J. 1994. American higher education: A history (p. 187). New York: St. Martin's Press.

Ma, A. 2015. You don't need to know how to code to make it in Silicon Valley. *LinkedIn*. https://blog.linkedin.com/2015/08/25/you-dont-need-to-know-how-to-code-to-make-it-in-silicon-valley (accessed October 9, 2017).

Maeda, J. 2013. Stem+ art= steam. *The STEAM Journal* 1(1):34.

Malavé, C. O., and K. L. Watson. 2000. The freshman integrated curriculum at Texas A&M University. http://www.ineer.org/Events/ICEE1998/Icee/papers/259.pdf.

Mangione, S., and L. Z. Nieman. 1997. Cardiac auscultatory skills of internal medicine and family practice trainees: a comparison of diagnostic proficiency. *JAMA* 278(9):717-722.

Mangione, S., and L. Z. Nieman. 1999. Pulmonary auscultatory skills during training in internal medicine and family practice. *American Journal of Respiratory and Critical Care Medicine,* 159(4):1119-1124.

Mansilla, V. B., I. Feller, and H. Gardner. 2006. Quality assessment in interdisciplinary research and education. *Research Evaluation* 15(1):69-74.

Margolis, J., A. Fisher, and F. Miller. 2000. The anatomy of interest: Women in undergraduate computer science. *Women's Studies Quarterly* 28(1/2):104-127.

Martin, P. E., and B. R. Umberger. 2003. Trends in interdisciplinary and integrative graduate training: An NSF IGERT example. *Quest* 55(1):86-94.

Mayhew, M. J., A. N. Rockenbach, N. A. Bowman, T. A. Seifert, G. C. Wolniak, E. T. Pascarella, and P. T. Terenzini. 2016. *How College Affects Students: 21st Century Evidence That Higher Education Works, Volume 3.* Hoboken, NJ: Jossey Bass.

Mehr, S.A., A. Schachner, R. C. Katz, and E. S. Spelke. Two randomized trials provide no consistent evidence for nonmusical cognitive benefits of brief preschool music enrichment. *PLoS One* 8(12):e82007. doi: 10.1371/journal.pone.0082007.eCollection 2013.

Meyer, J., and R. Land. 2003. *Threshold Concepts and Troublesome Knowledge: Linkages to Ways of Thinking and Practising within the Disciplines.* Edinburgh: University of Edinburgh.

Meyer, S. R., V. R. Levesque, K. H. Bieluch, M. L. Johnson, B. McGreavy, S. Dreyer, and H. Smith. 2016. Sustainability science graduate students as boundary spanners. *Journal of Environmental Studies and Sciences* 6(2):344-353.

Meyers, F. J., M. D. Begg, M. Fleming, and C. Merchant. 2012. Strengthening the career development of clinical translational scientist trainees: A consensus statement of the clinical translational science award (CTSA) research education and career development committees. *Clinical and Translational Science* 5(2):132-137.

Mikic, B. 2014. *Design Thinking and the Liberal Arts: A Framework for Re-Imagining a Liberal Arts Education,* S. College, ed. Northampton, MA: Smith College.

Millar, M. M. 2013. Interdisciplinary research and the early career: The effect of interdisciplinary dissertation research on career placement and publication productivity of doctoral graduates in the sciences. *Research Policy* 42(5):1152-1164.

Miller, E., D. Balmer, M. N. Hermann, M. G. Graham, and R. Charon. 2014. Sounding narrative medicine: Studying students' professional identity development at columbia university college of physicians and surgeons. *Academic Medicine* 89(2):335.

Miller, T. R., A. Wiek, D. Sarewitz, J. Robinson, L. Olsson, D. Kriebel, and D. Loorbach. 2014. The future of sustainability science: A solutions-oriented research agenda. *Sustainability Science* 9 (2):239-246. https://doi.org/10.1007/s11625-013-0224-6.

Miller, P. K., D. M. Doberneck, and J. H. Schweitzer. 2011. Translating University Knowledge to the General Public at the Art/Science/Engagement Interface. http://ncsue.msu.edu/files/TranslatingUniversityKnowledgePoster.pdf (accessed October 3, 2017).

MIT Media Lab. n.d. *Media Lab Spinoff Companies.* Cambridge, MA.

Mohler, J. L. 2007. An instructional strategy for pictorial drawing.

Muller, D. 2014. Flexmed: A nontraditional admissions program at Icahn School of Medicine at Mount Sinai. http://journalofethics.ama-assn.org/2014/08/medu2-1408.html (accessed October 9, 2017).

Naghshineh, S., J. P. Hafler, A. R. Miller, M. A. Blanco, S. R. Lipsitz, R. P. Dubroff, S. Khoshbin, and J. T. Katz. 2008. Formal art observation training improves medical students visual diagnostic skills. *Journal of General Internal Medicine* 23(7):991-997.

National Academy of Sciences. 2008. Science, Evolution, and Creationism. Washington, DC: The National Academies Press. https://doi.org/10.17226/11876.

National Academies of Science, Engineering, and Medicine. 2006. Taxonomy of fields and their subfields. http://sites.nationalacademies.org/pga/resdoc/pga_044522 (accessed September 29, 2017, 2017).

National Academies of Sciences, Engineering, and Medicine. 2015. *Integrating Discovery-Based Research into the Undergraduate Curriculum: Report of a Convocation*. Washington, DC: The National Academies Press. https://doi.org/10.17226/21851.

National Academy of Engineering. 2008. *Changing the Conversation: Messages for Improving Public Understanding of Engineering*. Washington, DC: The National Academies Press. https://doi.org/10.17226/12187.

National Academy of Engineering. 2012. Infusing Real World Experiences into Engineering Education. Washington, DC: The National Academies Press. https://doi.org/10.17226/18184.

National Academy of Engineering and National Research Council. 2009. *Engineering in K-12 Education: Understanding the Status and Improving the Prospects*. Washington, DC: The National Academies Press. https://doi.org/10.17226/12635.

National Academy of Engineering and National Research Council. 2014. *STEM Integration in K-12 Education: Status, Prospects, and an Agenda for Research*. Washington, DC: The National Academies Press. https://doi.org/10.17226/18612.

National Endowment for the Arts. Creative Forces: NEA Military Healing Arts Network. https://www.arts.gov/partnerships/creative-forces (accessed October 10, 2017).

National Endowment for the Arts. 2017. *Industrial Design: A Competitive Edge for U.S. Manufacturing Success in the Global Economy*. Washington, DC: National Endowment for the Arts.

National Research Council. 1995. *Colleges of Agriculture at the Land Grant Universities: A Profile*. Washington, DC: National Academy Press. https://doi.org/10.17226/4980.

National Research Council. 2000. *How People Learn: Brain, Mind, Experience, and School: Expanded edition*. Washington, DC: National Academy Press. https://doi.org/10.17226/6160.

National Research Council. 2009. *A New Biology for the 21st Century*. Washington, DC: The National Academies Press. https://doi.org/10.17226/12764.

Newswander, L. K., and M. Borrego. 2009. Engagement in two interdisciplinary graduate programs. *Higher Education* 58(4):551-562.

Nieusma, D. 2011. Materializing nano equity: Lessons from design. Pp. 209-230 in *Nanotechnology and the Challenges of Equity, Equality and Development*, S. Cozzens and J. Wetmore, eds. Dordrecht, Netherlands: Springer.

Nussbaum, M. C. 1997. *Cultivating Humanity: A Classical Defense of Reform in Liberal Education*. Cambridge, MA: Harvard University Press.

Nussbaum, M. C. 2005. Liberal education and global responsibility. Theholds in Education 31(3/4): 15.

Olds, B. M., and R. L. Miller. 2004. The effect of a first-year integrated engineering curriculum on graduation rates and student satisfaction: A longitudinal study. *Journal of Engineering Education* 93(1):23-35.

Ousager, J., and H. Johannessen. 2010. Humanities in undergraduate medical education: A literature review. *Academic Medicine Academic Medicine* 85(6):988-998.

Parthasarathy, S. 2016. Grassroots Innovation Systems for the Post-Carbon World: Promoting Economic Democracy, Environmental Sustainability, and the Public Interest. Brook. L. Rev., 82, 761.

PCAST (President's Council of Advisors on Science and Technology). 2012. Report to the President: Engage to Excel: Producing One Million Additional College Graduates with Degrees in Science, Technology, Engineering, and Mathematics. https://obamawhitehouse.archives.gov/sites/default/files/microsites/ostp/pcast-engage-to-excel-final_2-25-12.pdf (accessed October 10, 2017).

Pellico, L. H., T. C. Duffy, K. P. Fennie, and K. A. Swan. 2012. Looking is not seeing and listening is not hearing: Effect of an intervention to enhance auditory skills of graduate-entry nursing students. Nursing education perspectives 33(4):234-239.

Perry, M., N. Maffulli, S. Willson, and D. 2011. The effectiveness of arts-based interventions in medical education: a literature review. *Medical Education* 45(2):141-148.

Pesic, P. 2014. *Music and the Making of Modern Science*. Cambridge: MIT Press.

Peters, A. S., R. Greenberger-Rosovsky, C. Crowder, S. D. Block, and G. T. Moore. 2000. Long-term Outcomes of the New Pathway Program at Harvard Medical School: A Randomized Controlled Trial. Academic Medicine 75(5): 470-479.

Pew Research Center. 2011. Is College Worth It? College Presidents, Public Assess Value, Quality and Mission of Higher Education. Washington, DC: Pew Research Center. http://www.pewsocialtrends.org/files/2011/05/higher-ed-report.pdf (accessed October 10, 2017).

PhysOrg. 2010. Avant-garde music offers a gateway to artificial intelligence. https://phys.org/news/2010-12-avant-garde-music-gateway-artificial-intelligence.html (accessed October 10, 2017).

Pintrich, P. R., D. A. F. Smith, T. Garcia, and W. J. McKeachie. 1991. *A Manual for the Use of the Motivated Strategies for Learning Questionnaire (MSLQ)*. National Center for Research to Improve Postsecondary Teaching and Learning, Ann Arbor, MI.

Pollack, A. E., and D. L. Korol. 2013. The use of haiku to convey complex concepts in neuroscience. *Journal of Undergraduate Neuroscience Education: JUNE* 12(1):42-48.

Posselt, J. R., K. A. Reyes, K. E. Slay, A. Kamimura, and K. B. Porter. 2017. Equity efforts as boundary work: How symbolic and social boundaries shape access and inclusion in graduate education. *Teachers College Record* 119(10).

Rekers, J. V., and T. Hansen. 2014. Interdisciplinary research and geography: Overcoming barriers through proximity. Science and Public Policy 42(2): 242-254.

Rhode Island School of Design. 2016. Maharam STEAM fellows. https://risdmaharamfellows.com/ (accessed September 28, 2017).

Rhoten, D., E. O'Connor, and E. J. Hackett. 2009. The act of collaborative creation and the art of integrative creativity: Originality, disciplinarity and interdisciplinarity. *Thesis Eleven* 96(1):83-108.

Rhoten, D., and A. Parker. 2004. Risks and rewards of an interdisciplinary research path. *Science* 17(306)5704:2046.

Riley, D. 2015. Facepalms and cringes: Liberal education misapprehended. *Engineering Studies* 7(2-3):138-141.

Roche, M. W. 2013. The landscape of the liberal arts. *New Directions for Community Colleges* 163:3-10.

Root-Bernstein, R., L. Allen, L. Beach, R. Bhadula, J. Fast, C. Hosey, B. Kremkow, J. Lapp, K. Lonc, and K. Pawelec. 2008. Arts foster success: Comparison of Nobel prizewinners, Royal Society, National Academy, and Sigma xi members. *Journal of Psychology of Science and Technology* 1(2):51-63.

Root-Bernstein, R., and M. Root-Bernstein. 1999. Sparks of genius: The thirteen thinking tools of the world's most creative people. Boston and New York: Houghton Mifflin.

Saint-Louis, N., N. Seth, and K. S. Fuller. 2015. Curriculum integration: The experience of three founding faculty at a new community college. *International Journal of Teaching and Learning in Higher Education* 27(3):423-433.

Sands, S. A., P. Stanley, and R. Charon. 2008. Pediatric narrative oncology: Interprofessional training to promote empathy, build teams, and prevent burnout. *Journal of Supportive Oncology* 6(7):307-312.

Sapsed, J., and P. Nightingale. 2013. The Brighton fuse report. Brighton: Brighton Fuse.

Savilonis, B., D. Spanagel, and K. Wobbe. 2010. Engaging students with great problems. *ASEE Annual Conference and Exposition, Conference Proceedings.*

Schellnhuber, H.-J. 2010. Global sustainability: A nobel cause. Paper read at Interdisciplinary Nobel Laureate Symposium on Global, Sustainability, 2010, Cambridge.

Scholl, V. M., M. M. Iafrati, D. L. Long, and J. J. Pow. 2014. Motion picture science: A fully integrated fine arts/stem degree program. Integrated STEM Education Conference, Princeton, NJ. https://ieeexplore.ieee.org/document/6891049/.

Science Council. Our definition of science. http://sciencecouncil.org/about-science/our-definition-of-science/ (accessed September 29, 2017).

Secord, J. A. 2004. Knowledge in transit. *Isis* 95(4):654-672.

Settle, A., D. Goldberg, and V. Barr. 2013. Beyond computer science: Computational thinking across disciplines. Pp. 311-312 in *ITiCSE '13: Proceedings of the 18th ACM Conference on Innovation and Technology in Computer Science Education.*

Shapiro, J., and L. Rucker. 2003. Can poetry make better doctors? Teaching the humanities and arts to medical students and residents at the university of California, Irvine, college of medicine. *Academic Medicine* 78:953-957.

Shapiro, J., L. Rucker, and J. Beck. 2006. Training the clinical eye and mind: using the arts to develop medical students' observational and pattern recognition skills. *Medical Education* 40(3):263-268.

Shen, J., S. Jiang, and O. L. Liu. 2015. Reconceptualizing a college science learning experience in the new digital era: A review of literature. Pp. 61-79 in *Emerging Technologies for STEAM Education: Full Steam Ahead,* X. Ge, D. Ifenthaler, and J. Spector, eds. Cham, Switzerland: Springer.

Skubikowski, K. M., C. Wright, and R. Graf. 2010. *Social Justice Education: Inviting Faculty to Transform Their Institutions*. Sterling, VA, and London: Stylus.

Snow, C. P. 1959. *Two Cultures*. Washington, DC: American Association for the Advancement of Science.

Snyder, T. D., C. de Brey, and S. A. Dillow. 2016. Digest of Education Statistics 2014, NCES 2016-006. National Center for Education Statistics.

Society for Music Perception and cognition. Resources. http://www.musicperception.org/smpc-resources.html (accessed October 10, 2017).

Somerson, R. 2013. The art of critical making: An introduction. Pp. 19-31 in *The Art of Critical Making: Rhode Island School of Design on Creative Practice*. Wiley.

Sorby, S. A., and B. J. Baartmans. 1996. A Course for the Development of 3-D Spatial VisualizationSkills. *Engineering Design Graphics Journal* 60(1):13-20.

Sorby, S. A., and B. J. Baartmans. 2000. The Development and Assessment of a Course for Enhancing the 3-D Spatial Visualization Skills of First Year Engineering Students. *Journal of Engineering Education* 89(3):301-307.

Sorby, S. A. 2009a. Educational research in developing 3-D spatial skills for engineering students. *International Journal of Science Education*, 31(3):459-480.

Sorby, S. A. 2009b. Assessment of a" new and improved" course for the development of 3-D spatial skills. *Engineering Design Graphics Journal*, 69(3).

Sousa, D. A., and T. Pilecki. 2013. *From STEM to STEAM: Using Brain-Compatible Strategies to Integrate the Arts*. Thousand Oaks, CA: Corwin.

Spector, J. M. 2015. Education, training, competencies, curricula and technology. Pp. 3-14 in *Emerging Technologies for STEAM Education: Full Steam Ahead,* X. Ge, D. Ifenthaler, and J. Spector, eds. Cham, Switzerland: Springer.

Spiegel, D., C. Malchiodi, A. Backos, and K. Collie, K. 2006. Art therapy for combat-related PTSD: Recommendations for research and practice. *Art Therapy* 23(4):157-164.

Stanford Medicine. Scholarly concentration: Biomedical ethics and medical humanities. http://med.stanford.edu/bemh.html (accessed October 10, 2017).

STEAM. 2016. About us. http://steamwith.us/ (accessed September 28, 2017).

STEM Innovation Task Force. 2014. *STEM 2.0—an Imperative for Our Future Workforce.* Washington, DC: STEMconnector.

STEM Innovation Task Force. 2015. *Focus on Employability Skills for STEM Workers Points to Experiential Learning.* Washington, DC: STEMconnector.

Stentoft, D. 2017. From saying to doing interdisciplinary learning: Is problem-based learning the answer? *Active Learning in Higher Education* 18(1):51-61.

Stephan, K. D. 2002. All this and engineering too: A history of accreditation requirements. *IEEE Technology and Society Magazine* 21(3).

Stolk, J., and R. Martello. 2015. Can disciplinary integration promote students' lifelong learning attitudes and skills in project-based engineering courses. *International Journal of Engineering Education* 31(1):434-449.

Strategic National Arts Alumni Project. 2014. Making It Work: The Education and Employment of Recent Arts Graduates. http://snaap.indiana.edu/pdf/2014/SNAAP_AR_2014.pdf (accessed October 10, 2017).

Strengers, Y. A. A. 2014. Interdisciplinarity and industry collaboration in doctoral candidature: Tensions within and between discourses. *Studies in Higher Education* 39(4):546-559.

Suzuki, J. 2015. *Constitutional Calculus: The Math of Justice and the Myth of Common Sense.* Baltimore, MD: Johns Hopkins University Press.

Tanner Boyd, I. J., K. Van Sickle, W. Schwesinger, J. Michalek, and J. Bingener 2008. Music experience influences laparoscopic skills performance. *Journal of the Society of Laparoendoscopic Surgeons* 12(3):292.

The Evergreen State College. n.d. Fast facts. https://evergreen.edu/about/facts (accessed October 4, 2017).

The Lincoln Project. 2015. Public research universities: Changes in state funding. https://www.amacad.org/content/publications/publication.aspx?d=21942 (accessed October 10, 2017).

The Teagle Foundation. 2017. Double majors and creativity: Influences, interactions, and impacts. http://www.teaglefoundation.org/Grants-Initiatives/Grants-Database/Grants/Special-Project/Double-Majors-and-Creativity-Influences,-Interact (accessed October 10, 2017).

Thigpen, L., E. Glakpe, G. Gomes, and T. McCloud. 2004. A model for teaching multidisciplinary capstone design in mechanical engineering. *Proceedings—Frontiers in Education Conference* 3(Conf 34):S2G-1-S2G-6.

Tillotson, M. L. (1984). Effect of instruction in spatial visualization on spatial abilities and mathematical problem solving. Gainesville: University of Florida.

Tinto, V. 2003. Learning better together: The impact of learning communities on student success. *Higher Education monograph series* 1(8):1-8.

Tobias, S. 1993. Why poets just don't get it in the physics classroom: Stalking the second tier in the sciences. *NACADA Journal* 13(2):42-44.

Tobias, S. 2016. Revisiting the New Liberal Arts Initiative, 1980-1990. https://www.asee.org/engineering-enhanced-liberal-education-project/background/new-liberal-arts-initiative (accessed October 3, 2017).

Tsui, L. 2007. Effective strategies to increase diversity in stem fields: A review of the research literature. *The Journal of Negro Education* 555-581.

Turner, J., and P. Bernard. 1993. The 'German model' and the graduate school: The University of Michigan and the origin myth of the American university. *History of Higher Education Annual* 13.

Union College. CSC-10x: Union college's unique introduction to CS. http://cs.union.edu/intro/ (accessed October 10, 2017).

United Nations Development Programme. 2016. *Human Development Report 2016: Human Development for Everyone*. New York: United Nations.

Uttal, D. H., and C. A. Cohen. 2012. Spatial thinking and STEM education: When, why, and how?. Pp. 147-181 in Psychology of Learning and Motivation (vol. 57). Cambridge, MA: Academic Press.

van de Ven, A. L., M. H. Shann, and S. Sridhar. 2014. Essential components of a successful doctoral program in nanomedicine. *International Journal of Nanomedicine* 10:23-30.

Vaughn, L. 2012. *Bioethics: Principles, Issues, and Cases*. New York: Oxford University Press.

Vaz, R., and P. Quinn. 2014. Long term impacts of off-campus project work on student learning and development. Paper read at Frontiers in Education Conference (FIE), 2014 IEEE.

Wachtler, C., S. Lundin, and M. Troein. 2006. Humanities for medical students? A qualitative study of a medical humanities curriculum in a medical school program. *BMC Medical Education* 6(1):16.

Wallace, D., B. Vuksanovich, and K. Carlile. 2010. Work in progress–building up steam–exploring a comprehensive strategic partnership between stem and the arts. Paper read at proceedings of ASEE 2010 North Central Sectional Conference, Pittsburgh, PA.

Wang, K. 2015. Study on the careers of MIT Mechanical Engineering undergraduate alumni (Doctoral dissertation, Massachusetts Institute of Technology).

Watts, R. J., and O. Guessous. 2006. Civil Rights Activists in the Information Age: The Development of Math Literacy Workers. Circle working paper 50. The Center for Information and Research on Civic Learning and Engagement (CIRCLE). University of Maryland, School of Public Policy.

Westerhaus, M., A. Finnegan, M. Haidar, A. Kleinman, J. Mukherjee, and P. Farmer. 2015. The necessity of social medicine in medical education. *Academic Medicine* 90(5):565-568.

Willson, V. L., T. Monogue, and C. Malave. 1995. First year comparative evaluation of the Texas A&M freshman integrated engineering program. Paper read at Frontiers in Education Conference, 1995.

Wilson, J.D. 2002. Models of Centralized and Decentralized Budgeting within Universities. Michigan State University. https://www.ilr.cornell.edu/sites/ilr.cornell.edu/files/Models%20of%20Centralized%20and%20Decentralized%20Budgeting%20within%20Universities.pdf.

Wolken, D. J. n.d. *The American university and "the German model"—a transnational history*. https://mrperez.expressions.syr.edu/immigration/education/wolken/ (accessed October 4, 2017).

World Health Organization. Traditional medicine: Definitions. http://www.who.int/medicines/areas/traditional/definitions/en/ (accessed September 29, 2017).

Youngerman, E. 2017. Integrative learning in award-winning student writing: A grounded theory analysis: Unpublished dissertation.

Zakaria, F. 2015. *In Defense of a Liberal Education*. New York: WW Norton & Company.

Zerr, R. J. 2014. Congressional apportionment: Constitutional questions, data, and the first presidential veto—STIRS student case study. https://www.aacu.org/stirs/casestudies/zerr (accessed October 9, 2017).

Zhang, D., and J. Shen. 2015. Disciplinary foundations for solving interdisciplinary scientific problems. *International Journal of Science Education* 37(15):2555-2576.

Appendix I

Committee and Staff Biographies

COMMITTEE MEMBER BIOGRAPHIES

David J. Skorton *Chair,* (NAM) is the 13th Secretary of the Smithsonian. He assumed his position July 1, 2015. As Secretary, Skorton oversees 19 museums and galleries, 21 libraries, the National Zoo, and numerous research centers, including the Smithsonian Astrophysical Observatory, the Smithsonian Tropical Research Institute and the Smithsonian Environmental Research Center. Skorton, a cardiologist, is the first physician to serve as Secretary. He previously was the president of Cornell University, a position he held beginning in July 2006. He was also a professor in the Departments of Medicine and Pediatrics at Weill Cornell Medical College in New York City and in Cornell's Department of Biomedical Engineering at the College of Engineering. His research focus is congenital heart disease and cardiac imaging and image processing. Before becoming Cornell's president, Skorton was president of the University of Iowa from 2003 to 2006 and a member of its faculty for 26 years. An ardent and nationally recognized supporter of the arts and humanities, Skorton has called for a national dialogue to emphasize the importance of funding for these disciplines. He asserts that supporting the arts and humanities is a wise investment in the future of the country. Skorton is a strong proponent of business–university partnerships. He has been active in innovation and economic development at the state and national levels to bring business and universities together toward diversifying regional economies. He is past chair of the Business-Higher Education Forum, an independent, nonprofit organization of industry CEOs, leaders of colleges and universities, and foundation executives.

Skorton was a pioneer in applying computer analysis and processing techniques to cardiac imaging; he has published two major texts and numerous articles, reviews and book chapters on cardiac imaging and image processing. Skorton was elected to the Institute of Medicine of the National Academies (now the National Academy of Medicine) and is a fellow of the American Academy of Arts and Sciences and an elected member of the American Philosophical Society. A national leader in research ethics, he was the charter president of the Association for the Accreditation of Human Research Protection Programs Inc., the first group organized specifically to accredit human research protection programs.

Skorton is an avid amateur musician who plays the flute and the saxophone. He cohosted "As Night Falls—Latin Jazz," a weekly program on the University of Iowa's public FM radio station. He is currently a distinguished professor at Georgetown University. Skorton earned his bachelor's degree in psychology and his M.D. both from Northwestern University. He completed his medical residency and fellowship in cardiology at the University of California, Los Angeles.

Susan Albertine is senior scholar at the Association of American Colleges and Universities (AAC&U). Beginning in 2008, she directed the LEAP States Initiative at AAC&U, dedicated to strengthening liberal and general education in public institutions and state systems, and served as vice president, Office of Diversity, Equity, and Student Success. Albertine received her B.A. in English from Cornell University, her M.A. in English from SUNY Cortland, and her Ph.D. in English from the University of Chicago. She has served as co-leader of the Educated Citizen and Public Health initiative, a project co-sponsored by AAC&U, the Association for Prevention Teaching and Research, the Council of Colleges of Arts and Sciences, and the Association of Schools and Programs of Public Health. She was dean of the School of Culture and Society and professor of English at the College of New Jersey from 2002 to 2008. Previously, she served as vice provost for undergraduate studies, Temple University, and assistant to the provost, University of Pennsylvania. She has held faculty positions at the University of North Florida, St. Olaf College, and Susquehanna University. Her scholarship in American literature focuses on women's work during the growth phase of industrialization in the United States. A former public school teacher, Albertine has supported pre-school through college alignment through work with the Education Trust and the American Diploma Project. Her board service has included the Camden Academy Charter High School in Camden, New Jersey; the Advisory Board for the Delaware Study of Instructional Costs and Productivity—Faculty Study, University of Delaware; the Art Sanctuary, an African-American arts and letters organization based in Philadelphia; the Council of Colleges of Arts and Sciences; and the National Center for the First-Year Experience and Students in Transition.

Norman Augustine (NAS/NAE) is retired chairman and CEO of Lockheed Martin Corporation. Augustine attended Princeton University where he graduated with a B.S.E. in aeronautical engineering, magna cum laude, and an M.S.E. He was elected to Phi Beta Kappa, Tau Beta Pi, and Sigma Xi. In 1958 he joined the Douglas Aircraft Company in California where he worked as a research engineer, program manager, and chief engineer. Beginning in 1965, he served in the Office of the Secretary of Defense as assistant director of Defense Research and Engineering. He joined LTV Missiles and Space Company in 1970, serving as vice president, advanced programs and marketing. In 1973 he returned to the government as assistant secretary of the Army and in 1975 became under secretary of the Army, and later acting secretary of the Army. Joining Martin Marietta Corporation in 1977 as vice president of technical operations, he was elected as CEO in 1987 and chairman in 1988, having previously been president and COO. He served as president of Lockheed Martin Corporation upon the formation of that company in 1995, and became CEO later that year. He retired as chairman and CEO of Lockheed Martin in August 1997, at which time he became a lecturer with the rank of professor on the faculty of Princeton University where he served until July 1999. Augustine served on the President's Council of Advisors on Science and Technology under Democratic and Republican presidents and led the 1990 Advisory Committee on the Future of the U.S. Space Program and the 2005 National Academies commission that produced the landmark report, *Rising Above the Gathering Storm: Energizing and Employing America for a Brighter Economic Future.* Augustine has been presented the National Medal of Technology by the President of the United States and received the Joint Chiefs of Staff Distinguished Public Service Award. He has five times received the Department of Defense's highest civilian decoration, the Distinguished Service Medal. He is co-author of *The Defense Revolution and Shakespeare in Charge* and author of *Augustine's Laws and Augustine's Travels*. He holds 23 honorary degrees and was selected by Who's Who in America and the Library of Congress as one of "Fifty Great Americans" on the occasion of Who's Who's fiftieth anniversary. He has traveled in over 100 countries and stood on both the North and South Poles of the Earth.

Laurie Baefsky is executive director of ArtsEngine at the University of Michigan, and the Alliance for the Arts in Research Universities—or "a2ru." a2ru is a partnership of over 40 top-ranking research institutions committed to increasing the production, investment, and integration of arts and design practice, research, curricula, and programming in higher education.

In this capacity she works locally and nationally to support and strengthen the arts and integrative arts endeavors in higher education. Baefsky joined ArtsEngine and a2ru in August 2014 and has led arts integrated educational initiatives over the past 15 years. From 2007 to 2011 Baefsky established the USU ArtsBridge program at Utah State University, connecting university students with area schools and community organizations through arts-based interdisciplinary engaged learning initiatives. During this time, she also directed professional development efforts for northern Utah schools under the Beverley Taylor Sorenson Arts Learning Program, working with this state-funded interdisciplinary public schools initiative as lead coordinator of professional development for Utah's northern region schools.

Within grantmaking, Baefsky served as grants manager for the Utah Division of Arts and Museums in Salt Lake City, Utah where she oversaw the annual distribution of state and federal funding for individuals, organizations, communities and educators. A skilled grant writer herself, her efforts have resulted in over $5.3 million in arts funding through grants from federal, state and private sources. From 2014 to 2018 Baefsky served as principal investigator on two successive Andrew W. Mellon Foundation-supported research initiatives at the University of Michigan, including "SPARC—Supporting Practice in the Arts, Research and Curricula."From 2017 to 2018 she chaired a subcommittee reviewing the National Academy of Sciences' Cultural Programs. Locally in southeastern Michigan Baefsky is on the Board of Trustees for the Ann Arbor Summer Festival, and member of the Cultural Leaders Forum of the Ann Arbor Arts Alliance. Baefsky also worked in multiple capacities with the Virginia Arts Festival in southeastern Virginia—as their education director, director of development research, and as a creative consultant.

Baefsky began her career as a classical flutist and music educator, with degrees in flute performance from Stony Brook University, University of Michigan, and California State University, Fullerton. She has appeared with the Minnesota Orchestra, Utah Symphony, New World Symphony, and as a tenured member of the Virginia Symphony for 15 seasons. As a chamber artist, her performances have ranged from Symphony Space and Chamber Music Society of Lincoln Center, NYC to northeastern Morocco and Umbria, Italy. Baefsky previously taught music and interdisciplinary courses for eight summers at the Virginia Governor's School for the Humanities and Visual & Performing Arts at University of Richmond, and served as applied music faculty at multiple colleges and universities throughout southeastern Virginia.

Kristin Boudreau is the Paris Fletcher Distinguished Professor of Humanities and head of the department of humanities and arts at Worcester Polytechnic Institute. She has published three monographs, one edited collection, and dozens of essays on nineteenth-century American literature (some of them concerning pedagogical approaches), and has taught a wide array of literature courses for undergraduate and graduate students. In recent years she has turned her attention to transforming engineering education by contextualizing engineering challenges in their historical, cultural, geographic and political settings; in this capacity she has been involved in WPI's activities as a member of the KEEN network of engineering institutions. Recent publications in this field include "To See the World Anew: Learning Engineering Through a Humanistic Lens" in Engineering Studies 2015 and "A Game-Based Approach to Information Literacy and Engineering in Context" (with Laura Hanlan) in Proceedings of the Frontiers in Education Conference 2015, several papers on integrative teaching in proceedings of the American Society for Engineering Education. A classroom game she developed with students and colleagues at WPI, "Humanitarian Engineering Past and Present: Worcester's Sewage Problem at the Turn of the Twentieth Century" was chosen by the National Academy of Engineering as an "Exemplary Engineering Ethics Activity" that prepares students for "ethical practice, research, or leadership in engineering." She is currently working with students and colleagues to develop role-playing games based on the Flint, Michigan water crisis. She is also pursuing NSF-funded research on the environment in engineering programs for LGBTQ+ students.

Norman Bradburn is a senior fellow at NORC at the University of Chicago. He also serves as the Tiffany and Margaret Blake Distinguished Service Professor Emeritus in the faculties of the University of Chicago's Irving B. Harris Graduate School of Public Policy Studies, Department of Psychology, Booth School of Business and the College. He is a former provost of the University (1984-1989), chairman of the Department of Behavioral Sciences (1973-1979), and associate dean of the Division of the Social Sciences (1971-1973). From 2000 to 2004 he was the assistant director for social, behavioral, and economic sciences at the National Science Foundation. Associated with NORC since 1961, he has been its director and president of its Board of Trustees. Bradburn has been at the forefront in developing theory and practice in the field of sample survey research in the cultural sector. He co-directs the American Academy of Arts and Sciences' Humanities Indicators project and is principal investigator of the CPC's Cultural Infrastructure project. For the Humanities Indicators project he oversees the collation and analysis of data, the creation of reliable benchmarks to guide future analysis of the humanities, and the development of a consistent and sustainable means of updating the data. For the Cultural Infrastructure project he oversees the systematic measurement of recent building projects and their consequences, modeling levels of creativity and sustainability of individual arts organizations before and after building projects, and the overall cultural vibrancy and vitality of their cities or regions as a result. Bradburn is a fellow of the American Statistical Association, a fellow of the American Association for the Advancement of Science, and an elected member of the International Institute of Statistics. He was elected to the American Academy of Arts and Sciences in 1994. In 1996 he was named the first Wildenmann Guest Professor at the Zentrum for Umfragen, Methoden und Analyse in Mannheim, Germany. In 2004 he was given the Statistics Canada/American Statistical Association Waksberg Award in recognition of outstanding contributions to the theory and practice of survey methodology.

Al Bunshaft is senior vice president of global affairs and workforce of the future for the Americas of Dassault Systèmes. In this role he is responsible for new business development, academic sales and relationships, and coordination of the company's involvement in institutes and consortia. From 2013 to 2016, Bunshaft was president and CEO of Dassault Systèmes Government Solutions, the U.S. subsidiary he led the creation of, focused on serving U.S. government agencies. From 2010 until 2013, Bunshaft was Managing Director of Dassault Systèmes Americas. Prior to joining Dassault Systèmes in 2010, he had a 25-year career at IBM, holding various executive roles in R&D, strategic initiatives and general management. Bunshaft's expertise in 3D visualization, computer graphics and engineering-related software tools has been a special focus of his career. Beginning with his post-graduate work at the National Science Foundation's Center for Interactive Computer Graphics, he has led efforts to introduce new visualization-led processes into far-ranging industries. He is Dassault Systèmes' leading voice for science, technology, engineering and mathematics (STEM) education in the United States, where 3D visualization has been singled out by both the National Science Foundation and the National Institutes of Health as a transformative technology. From 2014 to 2017, Bunshaft was co-chair of the STEM Innovation Task Force of STEMconnector, a diverse consortium of organizations concerned with STEM education and the future of human capital in the United States. He was named by the organization as one of it's Top 100 CEO STEM leaders. He is a member of the U.S. Council on Competitiveness, which advances a pro-growth policy agenda to U.S. government representatives. In addition, he serves on the Massachusetts Governor's STEM Advisory Council and is a board member of the Massachusetts High Technology Council and the New York Hall of Science. He is a member of the President's Council of the Olin College of Engineering and the Advisory Board of the College of Engineering and Applied Sciences at the University at Albany, State University of New York (SUNY). He received his bachelor of science degree in computer science and mathematics from the University at Albany, SUNY. He has a master of science degree in computer engineering from Rensselaer Polytechnic Institute, where he was a researcher at the Center for Interactive Computer Graphics, a university-industry research center operated in conjunction with the National Science Foundation.

Gail Burd is the senior vice provost for academic affairs and a distinguished professor in molecular and cellular biology and cellular and molecular medicine at the University of Arizona. Burd was appointed the vice provost for academic affairs in August 2008. In this role, Burd works closely with campus leaders to coordinate programs that will advance the academic mission of the university and help colleges and departments develop and assess their academic degree programs. Burd's research program has focused on development and neural plasticity in the vertebrate olfactory system. She is the principal investigator on a successful research project on undergraduate STEM education funded by the Association of American Universities and the Leona and Harry A. Helmsley Charitable Trust, and her more recent research has centered around undergraduate science education. In prior administrative roles at the University of Arizona, Burd served as the associate dean for academic affairs in the college of science, the interim department head of molecular and cellular biology, and the associate department head of molecular and cellular biology. A fellow of the American Association for the Advancement of Science, she has chaired several committees for national professional organizations, served on numerous government panels for the National Institutes of Health and the National Science Foundation, and received awards for her undergraduate teaching.

Edward Derrick is an independent consultant on issues at the intersection of science, policy, and society, including research and innovation policy and management. He served as founding director of the American Association for the Advancement of Science (AAAS) Center of Science, Policy & Society Programs from July 2011 to April 2017 after serving as deputy director then acting director of the AAAS Science and Policy Programs. The Center of Science, Policy & Society Programs bridges the science and engineering community on one side, and policy makers and the interested public on the other. The programs address an array of topics in science and society, including the interplay of science with religion, law, and human rights; they also connect scientists and policy makers through programs in science and government, including the S&T Policy Fellowship program; and help improve the conduct of research through peer review and discussion of standards of responsible conduct. As chief program director, Derrick oversaw the programs, which grew under his leadership to a staff of more than 50 and an annual budget of more than $20 million, and served as a member of senior management at AAAS. Derrick first joined AAAS in 1998 as a member of the AAAS Research Competitiveness Program (RCP). RCP provides review and guidance to the science and innovation community. He became director of the program in January 2004, with responsibility for the development of new business and oversight of all aspects of the design and execution of projects. Derrick has participated directly in more than 60 RCP projects, having led committees to assist state and institutional planning for research, to review research centers and institutions and to advise state and international funds on major investments. Derrick is an honorary member of the National Academy of Inventors and an Alexander von Humboldt Fellow. He holds the Ph.D. from the University of Texas at Austin, with a dissertation in theoretical particle physics, and the B.S. from the Massachusetts Institute of Technology, with an undergraduate thesis in biophysics. Between degrees, he worked for Ontario Hydro in the Nuclear Studies and Safety Division.

E. Thomas Ewing is a history professor and associate dean of graduate studies, research, and diversity at the College of Liberal Arts and Human Sciences of Virginia Tech. His education includes a B.A. from Williams College and a Ph.D. in history from the University of Michigan. He teaches courses in Russian, European, Middle Eastern, and world history, gender/women's history, and historical methods. His publications include, as author, *Separate Schools: Gender, Policy, and Practice in the Postwar Soviet Union* (2010) and *The Teachers of Stalinism: Policy, Practice, and Power in Soviet Schools in the 1930s* (2002); as editor, *Revolution and Pedagogy: Transnational Perspectives on the Social Foundations of Education* (2005); and as co-editor, with David Hicks, *Education and the Great Depression: Lessons from a Global History* (2006). His articles on Stalinist education have been published in *Gender & History, American Educational Research Journal, Women's History Review, History of Education Quarterly, Russian Review*, and *The Journal of Women's History*. He has received funding from the National Endowment for the Humanities, the Spencer Foundation, and the National Council for Eurasian and East European Research.

J. Benjamin Hurlbut is associate professor of biology and society in the School of Life Sciences at Arizona State University. He is trained in science and technology studies with a focus on the history of the modern biomedical and life sciences. His research lies at the intersection of science and technology studies, bioethics, and political theory. Hurlbut studies the changing relationships between science, politics, and law in the governance of biomedical research and innovation in the 20th and 21st centuries, examining the interplay of science and technology with shifting notions of democracy, religious and moral pluralism, and public reason. He is the author of *Experiments in Democracy: Human Embryo Research and the Politics of Bioethics* (Columbia University Press, 2017) and co-editor of

Perfecting Human Futures: Transhuman Visions and Technological Imaginations (Springer, 2016). He received an A.B. in classics from Stanford University and a Ph.D. in the history of science from Harvard University. He was a postdoctoral fellow in the program on science, technology, and society at Harvard.

Pamela L. Jennings is a professor and head, Department of Art + Design, College of Design, North Carolina State University. She is also the CEO of CONSTRUKTS, Inc., a consumer electronics research company developing mixed-reality and wireless technologies for learning. Solution finding through computational thinking and creativity has been the catalyst of Jennings' policy, philanthropic, academic, research, and entrepreneurial projects. She has developed, taught, and advocated for integrative knowledge production across a range of institutions of creativity, learning, and research. She held the first joint professorship appointment between the School of Art and the Human Computer Interaction Institute at Carnegie Mellon University. There she developed an integrative curriculum that blended learning across new media arts, design, computer science, engineering, and human centered interaction. Three strategies informed her pedagogical framework: to celebrate and encourage a diversity of integrative skills and conceptual aptitudes; to support and grow self-efficacy in technology-based practices for women and underrepresented minorities in STEM; and to develop flexible approaches to address the broad continuum of technical and conceptual skills in her classroom. Jennings served as a National Science Foundation Program Officer in the Computer & Information Science & Engineering, directorate. There she led the CreativeIT program and co-led the Human Centered Computing program along with other synergistic NSF funding initiatives. The CreativeIT program funded high risk-high reward research at the nexus of creative cognition and creative practices with research in computer science, engineering, and STEM learning. She funded workshops that supported network building for the academic and professional art and technology field that resulted in the leadership identification and funding of the SEAD (Science, Engineering, Art, & Design) network. Jennings received her Ph.D. in Human Centered Systems Design in the Center for Advanced Inquiry in the Integrative Arts at the University of Plymouth School of Computing, Electronics, and Mathematics (United Kingdom); M.B.A. at the University of Michigan Ross School of Business; M.F.A. in computer art at the School of Visual Arts; M.A. in studio art in the International Center of Photography/New York University Program; and B.A. in psychology with vocal studies at Oberlin College.

Youngmoo Kim is director of the Expressive and Creative Interaction Technologies (ExCITe) Center and professor of electrical and computer engineering at Drexel University. His research group, the Music & Entertainment Technology Laboratory (MET-lab), focuses on the machine understanding of audio, particularly for music information retrieval. Other areas of active research at MET-lab include human-machine interfaces and robotics for expressive interaction, analysis-synthesis of sound, and K-12 outreach for engineering, science, and mathematics education. Youngmoo also has extensive experience in music performance, including 8 years as a member of the Tanglewood Festival Chorus, the chorus of the Boston Symphony Orchestra. He is a former music director of the Stanford Fleet Street Singers and has performed in productions at American Musical Theater of San Jose and SpeakEasy Stage Company (Boston). He is a member of Opera Philadelphia's newly formed American Repertoire Council. Youngmoo was named "Scientist of the Year" by the 2012 Philadelphia Geek Awards and was recently honored as a member of the Apple Distinguished Educator Class of 2013. He is recipient of Drexel's 2012 Christian R. and Mary F. Lindback Award for Distinguished Teaching. He co-chaired the 2008 International Conference on Music Information Retrieval hosted at Drexel and was invited by the National Academy of Engineering to co-organize the "Engineering and Music" session for the 2010 Frontiers of Engineering conference. His research is supported by the National Science Foundation and the John S. and James L. Knight Foundation.

Robert Martello is associate dean for curriculum and academic programs and professor of the history of science and technology at Olin College. He has chaired and initiated efforts that re-imagined Olin's faculty reappointment and promotion, institutional outreach, curricular innovation, and student assessment approaches. Martello's NSF-sponsored research, engineering education publications, and faculty development workshops explore connections between interdisciplinary integration, faculty teaming, student motivation, and project-based learning. He has delivered educational workshops for audiences around the world that include instructors and administrators at the K-12, community college, public, and private college levels. Martello implements his findings in experimental courses such as "The Stuff of History," "Six Microbes that Changed the World," "Paradigms, Predictions, and Joules," and "Chemistry in Context." A graduate of MIT's program in the History and Social Study of Science and Technology, he is the author of *Midnight Ride, Industrial Dawn: Paul Revere and the Growth of American Enterprise*, a study of Revere's multifaceted manufacturing career and of his many national impacts in pioneering America's transition into the industrial age. He is now researching Benjamin Franklin's printing and business endeavors, and he regularly lectures on Revere and Franklin, our "Founding Makers," for audiences of all ages and backgrounds.

Gunalan Nadarajan is dean and professor at the Penny W. Stamps School of Art and Design at the University of Michigan. His publications include *Ambulations* (2000), *Construction Site* (edited; 2004) and *Contemporary Art in Singapore* (co-authored; 2007), *Place Studies in Art, Media, Science and Technology: Historical Investigations on the Sites and Migration of Knowledge* (co-edited; 2009), *The Handbook of Visual Culture* (co-edited; 2012) and more than 100 book chapters, catalogue essays, academic articles, and reviews. His writings have also been translated into 16 languages. He has curated many international exhibitions including Ambulations (Singapore, 1999), 180KG (Jogjakarta, 2002), media_city (Seoul, 2002), Negotiating Spaces (Auckland, 2004), DenseLocal (Mexico City, 2009) and Displacements (Beijing, 2914). He was contributing curator for Documenta XI (Kassel, Germany, 2002) and the Singapore Biennale (2006) and served on the jury of a number of international exhibitions, including ISEA2004 (Helsinki / Talinn), transmediale 05 (Berlin), ISEA2006 (San Jose) and FutureEverything Festival (Manchester, 2009). He was artistic co-director of the Ogaki Biennale 2006, Japan, and artistic director of ISEA2008 (International Symposium on Electronic Art) in Singapore. He is active in the development of media arts internationally and has previously served on the Board of Directors of the Inter Society for Electronic Art and is on the Advisory Boards of the Database of Virtual Art (Austria), Cellsbutton Festival (Indonesia) and Arts Future Book series (UK). He currently serves on the International Advisory Board of the ArtScience Museum in Singapore. In 2013, he was elected to serve on the Board of Directors of the College Art Association. He has also served as an advisor on creative aspects of digital culture to the UNESCO and the Smithsonian Institution. He continues to work on a National Science Foundation funded initiative to develop a national network for collaborative research, education, and creative practice between sciences, engineering, arts, and design. He is a member of several professional associations including Special Interest Group in Graphics and Interactive Techniques (SIGGRAPH), Association for Computing Machinery (ACM), College Art Association, National Council of University Research Administrators, International Association of Aesthetics, International Association of Philosophy and Literature, and the American Association for the Advancement of Science. In 2004, he was elected a fellow of the Royal Society of Art. He has served in a variety of academic roles in teaching, academic administration, and research for over two decades. Prior to joining University of Michigan, he was vice provost for research and dean of graduate studies at the Maryland Institute College of Arts. He also had previous appointments as associate dean for research and graduate studies at the College of Arts and Architecture, Pennsylvania State University, and dean of visual arts at the Lasalle College of the Arts, Singapore.

Thomas F. Nelson Laird is an associate professor in the Higher Education and Student Affairs Program and director of the Center for Postsecondary Research at Indiana University Bloomington. Nelson Laird received a B.A. in mathematics from Gustavus Adolphus College (1995), an M.S. in mathematics from Michigan State University (1997), and a Ph.D. in higher education from the University of Michigan (2003). His work concentrates on improving teaching and learning at colleges and universities, with a special emphasis on the design, delivery, and effects of curricular experiences with diversity. He directs the activities of the Faculty Survey of Student Engagement, a companion project to the National Survey of Student Engagement, and the VALUE Institute, a collaboration with the Association of American Colleges and Universities. Author of many articles, chapters, and reports, Nelson Laird's work has appeared in key scholarly and practitioner publications. He also consults with institutions of higher education and related organizations on topics ranging from effective assessment practices to the inclusion of diversity into the curriculum.

Lynn Pasquerella is president of the Association of American Colleges and Universities. Assuming the presidency of the Association of American Colleges and Universities on July 1, 2016, throughout her career, Pasquerella has demonstrated a deep and abiding commitment to access to excellence in liberal education regardless of socioeconomic background. A philosopher, whose career has combined teaching and scholarship with local and global engagement, Pasquerella's presidency of Mount Holyoke College was marked by a robust strategic planning process, outreach to local, regional, and international constituencies, and a commitment to a vibrant campus community. A graduate of Quinebaug Valley Community College, Mount Holyoke College, and Brown University, Pasquerella joined the Department of Philosophy at the University of Rhode Island in 1985, rising rapidly through the ranks to the positions of vice provost for research, vice provost for academic affairs, and dean of the graduate school. In 2008, she was named provost at the University of Hartford. In 2010, her alma mater appointed her the eighteenth president of Mount Holyoke College. Pasquerella has written extensively on medical ethics, metaphysics, public policy, and the philosophy of law. At the core of her career is a strong commitment to liberal education and inclusive excellence, manifested in service as senator and vice president of Phi Beta Kappa; her role as host of Northeast Public Radio's "The Academic Minute"; and her public advocacy for access and affordability in higher education.

Suzanna Rose is founding associate provost for the Office to Advance Women, Equity and Diversity and professor of psychology and women's and gender studies at Florida International University (FIU). Rose also is the lead investigator for FIU's NSF ADVANCE Institutional Transformation grant that is aimed at improving the recruitment, promotion, and retention of women and underrepresented minority faculty at FIU. A key research project associated with the grant includes the development of an evidence-based Bystander Intervention program to reduce gender and race bias in faculty hiring, promotion, and retention. Her previous administrative roles included serving at FIU within the College of Arts and Sciences as executive director of the School of Integrated Science and Humanity, senior associate dean for the sciences, chair of psychology, and director of women's studies. Prior to that she served as women's studies director and professor of psychology at the University of Missouri-St. Louis. Rose has published extensively on issues related to gender, race, and sexual orientation, including professional networks, career development, leadership, friendship, and personal relationships. She has consulted with many universities both nationally and internationally concerning strategies for recruiting and retaining women faculty in science and engineering.

Bonnie Thornton Dill is dean of the University of Maryland College of Arts and Humanities and professor of women's studies. A pioneering scholar studying the intersections of race, class, and gender in the United States with an emphasis on African American women, work, and families, Thornton Dill's scholarship has been reprinted in numerous collections and edited volumes. Her recent publications include an edited collection of essays on intersectionality with Ruth Zambrana titled *Emerging Intersections: Race, Class, and Gender in Theory, Policy, and Practice* (Rutgers University Press, 2009), and numerous articles. Prior to assuming the position of dean, Thornton Dill chaired the Women's Studies Department for 8 years. In addition, she has worked with colleagues to found two research centers that have been national leaders in developing and disseminating the body of scholarship that has come to be known by the term "intersectionality." Today she holds the title of founding director for both the Center for Research on Women at the University of Memphis and the Consortium on Race, Gender and Ethnicity at the University of Maryland. She was president of the National Women's Studies Association (2010-2012) and prior to that was vice president of the American Sociological Association. Thornton Dill also serves as chair of the Advisory Board of Scholars for *Ms. Magazine*. Thornton Dill has won a number of prestigious awards including two awards for mentoring; the Jessie Bernard Award and the Distinguished Contributions to Teaching Award both given by the American Sociological Association; the Eastern Sociological Society's Robin Williams Jr. Distinguished Lectureship; and in 2009-2010, was appointed Stanley Kelley, Jr. Visiting Professor for Distinguished Teaching in the Department of Sociology at Princeton University. Her current research pulls together

her knowledge and experience as a teacher, mentor, and institution builder around issues of race/ethnicity, class, and gender in higher education to examine the experiences of historically underrepresented minority faculty in research universities, focusing specifically upon the impact of occupational stress on their physical and mental health and their career paths.

Laura Vosejpka is the founding dean of the College of Sciences and Liberal Arts at Kettering University (former General Motors Institute) in Flint, Michigan. Besides being responsible for the basic science majors at Kettering, she is also guiding an initiative to revamp the general education core for all engineers, providing a STEAM focused interdisciplinary approach to engineering education and she was a key player in bringing the Michigan Transfer Agreement for community college students to Kettering. Vosejpka came to Kettering after 6 years as a professor of physical science at Mid Michigan Community College (MMCC) in Harrison, Michigan. While at MMCC, she was an inaugural member of the Michigan Community College Association Leadership Academy, she served as chair of the General Education Committee and she was a co-leader for the college's participation in the Michigan Community College Association Guided Pathways Initiative aimed at improving retention and completion rates for community college students. She was responsible for the physics, pre-engineering, non-majors science and organic chemistry curricula. Prior to joining MMCC, Vosejpka served as the executive communications director for global R&D for the Dow Chemical Company and was responsible for providing internal and external executive communications support for the chief technology officer and the R&D leadership team. Her earlier work at Dow as an R&D specialist in core R&D was in the areas of biocatalysis and electroactive organic polymers (pLED). She is the author of six internal Dow research reports and was awarded the 2002 Chemical Sciences Technical Award for her work on pLED polydispersity and lifetime relationships.

Vosejpka held previous appointments at both Northwood University and Alma College. A passionate advocate for liberal arts education, Laura was a dual major in science and the humanities, graduating with Honors from The Ohio State University with B.A. degrees in both chemistry and english literature. She earned her Ph.D. in organic chemistry from the University of Wisconsin – Madison in 1989 and then spent 18 months as a postdoctoral research associate at the University of Maryland.

Lisa M. Wong is a musician, pediatrician, and past president of the Longwood Symphony Orchestra (LSO). She grew up in Honolulu, Hawaii where she attended Punahou School, an independent school centered on education, the arts and community service. She began the piano at age 4, violin at age 8, guitar at age 10, and viola at age 40. Wong graduated from

Harvard University in East Asian studies in 1979 and earned her M.D. from New York University's School of Medicine in 1983. After completing her pediatric residency at Massachusetts General Hospital in 1986, she joined Milton Pediatrics Associates and is an assistant clinical professor of pediatrics at Harvard Medical School. Wong is inspired by the work of Nobel Peace Prize laureate Dr. Albert Schweitzer, a humanitarian, theologian, musician, and physician. During her 20-year tenure as president of the LSO, was honored to work with remarkable leaders in health care and humanitarianism including Lachlan Forrow, Jackie Jenkins-Scott, Jim O'Connell, and Paul Farmer. Although she retired as president of the LSO in 2012, Wong continues her involvement with the orchestra as a violinist in the section. A passionate arts education advocate, Wong has worked closely with the New England Conservatory of Music's Preparatory School and traveled with NEC's Youth Philharmonic Orchestra to Brazil, Cuba, Guatemala, Panama, and Venezuela as a pediatric chaperone. Wong continues to be actively involved in El Sistema USA and has had the privilege of observing El Sistema in Venezuela several times over the past 10 years. Wong served as board member of Young Audiences of Massachusetts for more than 15 years and helped start Bring Back the Music (now renamed Making Music Matters), a program that revitalized in-class instrumental music instruction in the four Boston public elementary schools. In 2009, Wong was appointed to the Board of the Massachusetts Cultural Council by Governor Deval Patrick. In April 2010, Wong received the Community Pinnacle Award from Mattapan Community Health Center for LSO's pivotal role in their capital campaign to build a new neighborhood healthcare facility. Her first book *Scales to Scalpels: Doctors Who Practice the Healing Arts of Music and Medicine*, co-written with Robert Viagas, was published in April 2012 by Pegasus Books. It was released as a paperback in May 2013 and was recently translated into Chinese. The audiobook version will be released in early 2014.

STAFF BIOGRAPHIES

Ashley Bear is a program officer with the Board on Higher Education and Workforce at the National Academies of Sciences, Engineering, and Medicine. Before joining to the National Academies, Bear was a presidential management fellow with the National Science Foundation's (NSF) Division of Biological Infrastructure in the Directorate for Biological Sciences, where she managed a portfolio of mid-scale investments in scientific infrastructure and led analyses of the impact of NSF funding on the career trajectories of postdoctoral researchers. During her fellowship years, Bear also worked as a science policy officer for the State Department's Office of the Science and Technology Adviser to the Secretary of State, where she worked to promote science diplomacy and track emerging scientific trends with implications for foreign policy, managed programs to increase the scientific capacity of the State Department, and acted as the liaison to the Bureau of Western Hemisphere Affairs and the Bureau of East Asian and Pacific Affairs. Bear holds a Sc.B. in neuroscience from Brown University and a Ph.D. in ecology and evolutionary biology from Yale University. While working on her doctoral research one the developmental basis of courtship behavior in butterflies, Bear co-founded the Evolution Outreach Group, a volunteer organization composed of students and postdoctoral researchers that visit schools, museums, and camps in the greater New Haven, Connecticut area to teach K-12 students about evolution through hands-on activities and demonstrations. Bear is passionate about science outreach to the public and about promoting diversity and inclusion in science, technology, engineering, and mathematics.

Austen Applegate is a senior program assistant with the Board on Higher Education and Workforce (BHEW) at the National Academies of Sciences, Engineering, and Medicine. Prior to joining the National Academies he worked in a number of professional fields including international development, clinical research, and education. Applegate received his B.A. in psychology and sociology from Guilford College. There he developed his interest in social sciences and policy through his coursework in public health, health policy, behavioral medicine, qualitative and quantitative research methodology, race and gender disparities, and social science history. Applegate plans to pursue a Master in Public Health in the future.

Adriana Navia Courembis joined the Academies in January 2012 as part of the Finance Staff for the Policy and Global Affairs Division. At this position she collaborates with the financial management for the Board on Higher Education and Workforce, the Committee on Women in Science, Engineering and Medicine, the Science & Technology for Sustainability Program, the Committee on Human Rights, and the Board on Research Data and Information. Prior to the Academies, Courembis worked with the American Bar Association - Rule of Law Initiative as a Program Associate and Bay Management, LLC as an Accounts Payable Associate. Courembis holds a Bachelor in Arts in International Economics from American University.

Elizabeth Garbee was a Christine Mirzayan Science and Technology Policy Fellow at the National Academies during the spring of 2018. She earned her Ph.D. in science policy from the Consortium for Science Policy and Outcomes (CSPO) at Arizona State University, having first earned a bachelor's degree in astrophysics and classical civilizations from Oberlin College of Arts and Sciences. Her focus is higher education policy as it relates to science and technology, with additional expertise in risk perception, decision making, and public policy. Prior to joining the National Academies as a Mirzayan Fellow, she worked as a summer associate with the Science and Technology Policy Institute supporting tasks serving the Pentagon and the National Science Foundation.

Kellyann Jones-Jamtgaard is the career academy liaison at the Partnership for Regional Educational Preparation-Kansas City (PREP-KC), an education nonprofit that focuses on college and career preparation for urban school districts. Jones-Jamtgaard was a 2017 Christine Mirzayan Science and Technology Policy Fellow assigned to the Committee on Women in Science, Engineering, and Medicine at the National Academies of Sciences, Engineering, and Medicine. Appointed by Mayor Sly James, Jones-Jamtgaard currently serves as a commissioner on the Kansas City Health Commission, a group tasked with improving public health in Kansas City, Missouri, and co-chairs the Commission's Birth Outcomes subcommittee. Jones-Jamtgaard holds a B.S. in biology and Spanish from Duke University and a Ph.D. in microbiology from the University of Kansas Medical Center (KUMC). Her doctoral research focused on alterations in cellular trafficking during Hepatitis C virus infection. During graduate school, Jones-Jamtgaard was a member of the Committee for Postdocs and Students through the American Society for Cell Biology, co-chairing its career development subcommittee and serving as a liaison to the Public Policy and Minority Affairs committees. Jones-Jamtgaard is committed to improving science education and being an advocate for women in science and medicine. She was recently recognized with the naming of the Kellyann Jones-Jamtgaard Student Diversity Award at KUMC in her honor.

Jay B. Labov is senior advisor for education and communication for the National Academies of Sciences, Engineering, and Medicine. He has directed or contributed to 25 National Academies reports focusing on undergraduate education, teacher education, advanced study for high school students, K-8 education, and international education. He has served as director of committees on K-12 and undergraduate science education, the National Academies' Teacher Advisory Council, and was deputy director for the Academies' Center for Education. He directed a committee of the National Academy of Sciences and the Institute of Medicine that authored *Science, Evolution, and Creationism* and oversees the National Academy of Sciences' efforts to confront challenges to teaching evolution in the nation's public schools. He coordinates efforts at the National Academies to work with professional societies and with state academies of science on education issues. He also oversees work on improving education in the life sciences under the aegis of the Academy's Board on Life Sciences. Labov is an organismal biologist by training. Prior to accepting his position at the Academy in 1997, he spent 18 years on the biology faculty at Colby College (Maine). He is a Kellogg National Fellow, a fellow in Education of the American Association for the Advancement of Science, a Woodrow Wilson Visiting Fellow, and a 2013 recipient of the "Friend of Darwin" award from the National Center for Science Education. In 2013 he was elected to a 3-year term beginning in 2014 in which he served as chair-elect for 2014, chair for 2015 and past chair for 2016 of the Education Section of the American Association for the Advancement of Science. In 2014 he was named a Lifetime Honorary Member by the National Association of Biology Teachers, that organization's highest award and recognition. He received a National Academies Staff Award for Lifetime Achievement in December 2014 and was named by the Society for Integrative and Comparative Biology as the John A. Moore Lecturer for 2016.

Irene Ngun is a research associate with the Board on Higher Education and Workforce (BHEW) at the National Academies of Sciences, Engineering, and Medicine. She also serves as research associate for the Committee on Women in Science, Engineering, and Medicine (CWSEM), a standing committee of the National Academies. Before joining the National Academies she was a congressional intern for the U.S. House Committee on Science, Space, and Technology (Democratic Office) and served briefly in the office of Congresswoman Eddie Bernice Johnson of Texas (D-33). Ngun received her M.A. from Yonsei Graduate School of International Studies (Seoul, South Korea), where she developed her interest in science policy. She received her B.A. from Goshen College in biochemistry/molecular biology and global economics.

Thomas Rudin is the director of the Board on Higher Education and Workforce (BHEW) at the National Academies of Sciences, Engineering, and Medicine—a position he assumed in mid-August 2014. Prior to joining the Academies, Rudin served as senior vice president for career readiness and senior vice president for advocacy, government relations, and development at the College Board from 2006 to 2014. He was also vice president for government relations from 2004 to 2006 and executive director of grants planning and management from 1996 to 2004 at the College Board. Before joining the College Board, Rudin was a policy analyst at the National Institutes of Health in Bethesda, Maryland. In 1991, Rudin taught courses in U.S. public policy, human rights, and organizational management as a visiting instructor at the Middle East Technical University in Ankara, Turkey. In the early 1980s, he directed the work of the Governor's Task Force on Science and Technology for North Carolina Governor James B. Hunt, Jr., where he was involved in several new state initiatives, such as the North

Carolina Biotechnology Center and the North Carolina School of Science and Mathematics. He received a B.A. from Purdue University, and he holds master's degrees in public administration and in social work from the University of North Carolina at Chapel Hill.

J. D. Talasek is the director of Cultural Programs of the National Academy of Sciences (www.cpnas.org). Talasek is creator and moderator for a monthly salon called DASER (DC Art Science Evening Rendezvous) held at the NAS. He is currently on the faculty at Johns Hopkins University in the Museum Studies Master's Program. Additionally, Talasek serves on the Contemporary Art and Science Committee (CASC) at the Smithsonian's National Museum of Natural History. He is the art advisor for *Issues in Science and Technology Magazine* and is currently the Art and Design Advisor for the National Academies Keck Futures Initiative based in Irvine, California. Talasek is a board member of Leonardo/ International Society for Art Science and Technology and is chair of the Leonardo Education Arts Forum. He was the creator and organizer of two international online symposia (and coeditor of the subsequent published transcripts: *Visual Culture + Bioscience* (2009, DAP) and *Visual Culture + Evolution* (2010, DAP).

Appendix II

Statement of Task

An ad hoc committee overseen by the Board on Higher Education and Workforce (BHEW), in collaboration with units in PGA, NAE, IOM, and DBASSE, will produce a consensus report that examines the evidence behind the assertion that educational programs that mutually integrate learning experiences in the humanities and arts with science, technology, engineering, math, and medicine (STEMM) lead to improved educational and career outcomes for undergraduate and graduate students. In particular, the study will examine the following:

- Evidence regarding the value of integrating more STEMM curricula and labs into the academic programs of students majoring in the humanities and arts in order to understand the following: (1) how STEMM experiences provide important knowledge about the scientific understanding of the natural world and the characteristics of new technologies, knowledge that is essential for all citizens of a modern democracy; (2) how technology contributes essentially to sound decision making across all professional fields; and (3) how STEMM experiences develop the skills of scientific thinking (a type of critical thinking), innovation, and creativity that may complement and enrich the critical thinking and creativity skills developed by the arts and humanities.
- Evidence regarding the value of integrating curricula and experiences in the arts and humanities—including , history, literature, philosophy, culture, and religion—into college and university STEMM education programs, in order to understand whether and how

these experiences: (1) prepare STEMM students and workers to be more effective communicators, critical thinkers, problem-solvers and leaders; (2) prepare STEMM graduates to be more creative and effective scientists, engineers, technologists, and health care providers, particularly with respect to understanding the broad social and cultural impacts of applying knowledge to address challenges and opportunities in the workplace and in their communities; and (3) develop skills of critical thinking, innovation, and creativity that may complement and enrich the skills developed by STEMM fields.
- New models and good practices for mutual integration of the arts and humanities and STEMM fields at 2-year colleges, 4-year colleges, and graduate programs, drawing heavily on an analysis of programs that have been implemented at institutions of higher education.

The report will summarize the results of this examination and provide recommendations for all stakeholders to support appropriate endeavors to strengthen higher education initiatives in this area.

Appendix III

Meeting Agendas

COMMITTEE MEETING 1:

July 27-July 28, 2016
NAS Building Room 120
2101 Constitution Avenue NW, Washington, D.C.

Wednesday, July 27, 2016

CLOSED SESSION

2:00 PM Committee Introductions and Review of Statement of Task
David J. Skorton, Committee Chair
Ashley Bear, Study Director

OPEN SESSION

3:00 PM Committee hears from project sponsors (National Endowment for the Arts, National Endowment for the Humanities)

4:00 PM Committee discusses the goals of the study and broader questions, such as:

- What evidence exists on the impact of educational experiences that integrate the arts, humanities, and STEM?
- What kinds of integrated programs exist and which disciplines and sub-disciplines from the humanities, arts, and STEM are most typically integrated?
- How are the arts, humanities, and STEM distinct from each other? Are they really so different?

- Are there skills and competencies that are distinctly developed through the study of the arts, vs. the humanities, vs. STEM?

5:30 PM Committee hears input from audience members and guests

6:00 PM Reception in the Great Hall

Thursday, July 28, 2016

CLOSED SESSION

8:30 AM Continental Breakfast

9:00 AM Committee reflects on discussion from previous day
Committee begins to identify the key questions the study should address

OPEN SESSION

10:00 AM Presentation and Committee Discussion
Robert Root-Bernstein, Professor of Physiology, Michigan State University

10:45 AM Coffee Break

11:00 AM Panel Discussion
William "Bro" Adams, Chairman of the National Endowment for the Humanities
Richard Miller, President of Olin College of Engineering

12:00 PM Lunch

1:00 PM Committee hears input from audience members and guests

CLOSED SESSION

2:00 PM Committee Discussion of Presentations, possible sub-committee working group topics based on expertise and interest, topics and venues for regional workshops.

4:00 PM Adjourn

REGIONAL WORKSHOP 1

Le Laboratoire Cambridge
650 East Kendall Street
Cambridge, MA
October 13-October 14, 2016

Thursday, October 13, 2016

OPEN SESSION

12:00 PM Informal boxed lunch

1:00 PM Panel I: Models, Practices, Opportunities, and Challenges for Mutual Integration of the Arts, Humanities, and Engineering
Amy Banzaert, Lecturer in Engineering at Wellesley College
Rick Vaz, Director, Center for Project-Based Learning, Worcester Polytechnic Institute
Emma Smith Zbarsky, Associate Professor, Department of Applied Mathematics, Wentworth Institute of Technology

2:00 PM Presentation on concepts of integration in higher education
Kevin Hamilton, Professor and Senior Associate Dean in the College of Fine and Applied Arts at the University of Illinois, Urbana-Champaign

3:00 PM Informal committee discussion with Howard Gardner, John H. and Elisabeth A. Hobbs Professor of Cognition and Education at the Harvard Graduate School of Education

CLOSED SESSION

3:30 PM Committee closed session

5:30 PM Committee closed reception (CaféArt Science)

6:30 PM Committee closed working dinner

Friday, October 14, 2016

OPEN SESSION

8:00 AM Continental breakfast

8:30 AM Presentation on integrative teaching and learning in the scholarly literature
Matthew Mayhew, William Ray and Marie Adamson Flesher Professor of Educational Administration at The Ohio State University

9:30 AM Panel II: Models, Practices, Opportunities, and Challenges for Mutual Integration of the Arts, Humanities, and Technology
Ben Schmidt, Assistant Professor of History, Northeastern University
Rosalind Williams, Bern Dibner Professor of the History of Science and Technology, MIT
Bret Eynon, Historian and Associate Provost at LaGuardia Community College (CUNY)

10:30 AM Coffee Break

10:45 AM Panel III: Models, Practices, Opportunities, and Challenges for Mutual Integration of the Arts, Humanities, and Science
Dan Brabander, Professor of Geosciences, Wellesley College
Vandana Singh, Professor of Physics, STIRS Scholar, Framingham State University
Catherine Pride, STIRS Fellow, Associate Professor of Psychology, Middlesex Community College
Loren B. Byrne, Associate Professor of Biology and Environmental Science Coordinator, STIRS Scholar, Roger Williams University

12:00 PM Welcome remarks by David Edwards, Professor of the Practice of Idea Translation in the School of Engineering and Applied Sciences at Harvard University and Founder of Le Laboratiore followed by a musical performance of excerpts from Bach Goldberg Variations by Justin Lo, violin (HMS '17); Michael Wu, cello (HMS '18); and committee member Lisa Wong, viola

12:15 PM Lunch

1:00 PM Discussion of Institutional Barriers and Opportunities for Mutual Integration of the Arts, Humanities, Science, Technology, Engineering, Math, and Medicine
Bob Pura, President of Greenfield Community College
Lee Pelton, President of Emerson College
Helen Drinan, President of Simmons College
Pam Eddinger, President of Bunker Hill Community College

2:00 PM Panel V: Models, Practices, Opportunities, and Challenges for Mutual Integration of the Arts, Humanities, and Medicine
Joel Katz, Director, Internal Medicine Residency Program, Harvard Medical School
Michelle Morse, Deputy Chief Medical Officer, Partners In Health, Founding Co-Director, EqualHealth, Assistant Program Director, Brigham and Women's Internal Medicine Residency
Rita Charon, Director of the Program in Narrative Medicine at the Columbia University
Ed Hundert, Dean for Medical Education, Harvard Medical School

3:15 PM Closed committee discussion

5:00 PM Adjourn

COMMITTEE MEETING 2:

NAS Building Room 120
2101 Constitution Avenue NW, Washington, D.C.
February 8-9, 2017

Wednesday, February 8, 2017

OPEN SESSION

8:30 AM Continental breakfast

9:00 AM	Presentation on review of the published research on the integration of the humanities, arts, and STEM in undergraduate education followed by discussion with the committee *Matthew Mayhew, William Ray and Marie Adamson Flesher Professor of Educational Administration at The Ohio State University*
10:00 AM	Coffee break
10:15 AM	Overview of the literature on the integration of the humanities, arts, and medicine *Kellyann Jones-Jamtgaard, 2017 Mirzayan Fellow*
10:45 AM	Results of research effort to capture and categorize integrative models and practices in higher education *Irene Ngun, Research Associate, National Academies of Sciences, Engineering, and Medicine*
11:00 AM	Committee discussion of the results of the literature review and the research on integrative models and practices
11:30 AM	Committee hears input from the public
12:00 PM	Lunch

CLOSED SESSION

1:00 PM	Overview of the report writing and review process *Marilyn Baker, Director for Reports and Communications*
2:00 PM	Closed discussion of the report outline in subcommittee working groups
6:00 PM	Closed Committee working dinner at Ris (2275 L St NW, Washington, DC 20037)

Thursday, February 9, 2017

CLOSED SESSION

8:30 AM Continental Breakfast

9:00 AM Closed Committee reflection on discussion from previous day and the outline

10:30 AM Coffee Break

10:45 AM Committee discussion on next steps

12:00 PM Adjourn

REGIONAL WORKSHOP 2:

April 6-7, 2017
Arizona State University
Life Sciences Building Room C202
C Wing 401 E. Tyler Mall
Tempe, Arizona

Thursday, April 6, 2017

OPEN SESSION

8:30 AM Continental breakfast

9:00 AM Introduction and welcome from the National Academies
Ashley Bear, Study Director

9:15 AM Session I: The Faculty and Practitioner's Perspective on Integration (Panel I). Moderated by Guna Nadarajan, Dean and Professor at the Penny W. Stamps School of Art and Design at the University of Michigan (member of the study committee)
David Guston, Director for the Future of Innovation in Society, Arizona State University
David Weaver, Professor of Physics, Estrella Mountain Community College

	Sha Xin Wei, Professor and Director, School of Arts, Media and Engineering, Arizona State University *Liz Lerman, Founder of Liz Lerman Dance Exchange and Institute, Professor at the Herberger Institute for Design and the Arts, Arizona State University*
10:30 AM	Coffee Break
10:45 AM	Session II: The Faculty and Practitioner's Perspective on Integration (Panel II). Moderated by Pamela Jennings, President and CEO of CONSTRUKTS, Inc. (member of the study committee) *Andrea Polli, Professor of Art and Ecology, University of New Mexico* *JoAnn Kuchera-Morin, Professor of Media Arts & Technology and Music, University of California* *Ed Finn, Assistant Professor, School of Arts, Media + Engineering, Arizona State University* *Fritz Breithaupt, Professor of Germanic Studies, Indiana University Bloomington*
12:00 PM	Lunch
1:00 PM	Session III: The Academic Administrator's Perspective on Integration. Moderated by Bonnie Thornton Dill, Dean and Professor of Women's Studies, University of Maryland, College of Arts and Humanities (member of the study committee) *Raymond Tymas-Jones, Associate Vice President for the Arts and Dean, University of Utah College of Fine Arts* *Joaquin Ruiz, Executive Dean, College of Letters, Arts, and Science, University of Arizona* *Maria Hesse, Vice Provost for Academic Partnerships, Arizona State University*
2:00 PM	Session IV: Perspectives on Integration from Emerging and Established Thought Leaders in Innovation. Moderated by Laurie Baefsky, Executive Director, ArtsEngine and the Alliance for the Arts in Research Universities (a2ru) (member of the study committee) *Jim Spohrer, Director of Cognitive OpenTech, IBM Research – Almaden*

Dan Nathan-Roberts, Assistant Professor in Industrial and Systems Engineering, San Jose State University
Ethan Eagle, Assistant Professor of Mechanical Engineering, Wayne State University
Aileen Huang-Saad, Assistant Professor, Biomedical Engineering, Entrepreneurship and Engineering Education

3:15 PM Coffee Break

3:30 PM Session V: The ASU Student's Perspective on Integration. Moderated by Ben Hurlbut, Assistant Professor of Biology and Society in the School of Life Sciences, Arizona State University (member of the study committee)
Tess Doezema, Doctoral Student in Human and Social Dimensions of Science and Technology
Anna Guerrero, Undergraduate Biology Major and Artist
Cecilia Chou, Masters Student in Biology and Society
Matt Contursi, Undergraduate Double Major in English and Molecular Biosciences and Biotechnologies

4:45 PM Committee hears comments from the audience

5:45 PM Closing Remarks from the National Academies
Ashley Bear, Study Director

6:00 PM-7:00 PM Reception

Friday, April 7, 2017

CLOSED SESSION

8:30 AM Continental breakfast

9:00 AM Committee reflection on the workshop presentations

9:45 AM Lessons learned from the dissemination process "Rising Above the Gathering Storm" report
Norman Augustine, Committee Member

10:15 AM Committee breaks up into writing groups to discuss drafts

11:15 AM Full committee discussion of next steps

12:00 PM Adjourn

COMMITTEE MEETING 3:

July 13-14, 2017
Keck Building Room 201
500 5th Street NW, Washington, D.C.
Thursday, July 13, 2017

CLOSED SESSION

8:30 PM Continental breakfast

9:00 AM Full committee discussion about the report

10:00 AM Coffee break

10:30 AM Full committee discussion of report continued

11:45 AM Lunch

OPEN SESSION

12:00 PM Presentation on "The Liberal Sciences: Questioning and Learning from History"
Dr. Peter Pesic, Director of the Science Institute and Musician-in-Residence at St. John's College, Santa Fe and author of the book Music and the Making of Modern Science

CLOSED SESSION

1:00 PM Committee writing group breakout sessions

3:00 PM Writing groups report out to the full committee and offer initial suggestions for findings and recommendations

6:00 PM Closed Committee working dinner at Jaleo (480 7th St. NW)

Friday, July 14, 2017

8:30 AM Continental Breakfast

OPEN SESSION

9:00 AM Virtual presentation on the Value of Cognitive Diversity on Complex Tasks
Dr. Scott Page, Leonid Hurwicz Collegiate Professor of Complex Systems, Political Science, and Economics at the University of Michigan

10:00 AM Coffee break

CLOSED SESSION

10:15 AM Committee discussion about the report

12:15 PM Lunch

1:30 PM Committee discussion on next steps

3:30 PM Adjourn

COMMITTEE MEETING 4:

October 19-20, 2017
Keck Building
500 5th Street NW, Washington, D.C.

Thursday, October 19, 2017

CLOSED SESSION

8:30 PM Continental breakfast and thank you remarks from Bill O'Brien from the National Endowment from the Arts

9:00 AM Discussion on the nature of findings and recommendations in National Academies reports
David J. Skorton, Committee Chair
Ashley Bear, Study Director

9:15 AM Committee discussion about the final report
Ashley Bear, Study Director

10:00 AM	Coffee break
10:15 AM	Continued discussion about the final report
12:00 PM	Lunch
1:00 PM	Committee discussion on Integration at the Graduate Level
3:00 PM	Coffee Break

OPEN SESSION

3:30 PM	Discussion of dissemination strategy
5:00 PM	Reception
6:30 PM	Closed Committee working dinner

Friday, October 20, 2017

CLOSED SESSION

8:30 AM	Continental Breakfast
9:00 AM	Committee feedback for National Academies staff via short anonymous survey
9:15 AM	Discussion of the review process and recommendations from committee member
10:30 AM	Coffee Break
10:45 AM	Final committee discussion about the report
12:00 PM	Lunch
1:00 PM	Committee sign-off on the report and recommendations